CATALOGUE
OF FOSSIL HOMINIDS

PART I: AFRICA
(SECOND EDITION)

CATALOGUE OF FOSSIL HOMINIDS

PART I: AFRICA
(SECOND EDITION)

Editors:

KENNETH PAGE OAKLEY

BERNARD GRANT CAMPBELL

AND

THEYA IVITSKY MOLLESON

Publication No. 661
ISBN 0 565 05661 1

BMNH/B.17.76/1M/9.77

Printed in England by Staples Printers Limited at The George Press, Kettering Northamptonshire

CONTENTS

	Page
INTRODUCTION	ix
ALGERIA	1
CHAD	7
EGYPT	9
ETHIOPIA	17
KENYA	31
LIBYA	67
MALAWI	69
MALI	73
MAURITANIA	75
MOÇAMBIQUE	77
MOROCCO	79
NIGER	89
NIGERIA	91
RHODESIA	93
SOUTH AFRICA	95
SOUTH WEST AFRICA	153
SUDAN	155
TANZANIA	161
TUNISIA	186
ZAIRE	187
ZAMBIA	191
PLATES	201
INDEX	215

LIST OF TABLES

		Page
TABLE 1	Stratigraphic relationship of hominids from Hadar Formation, Ethiopia	19
TABLE 2	Stratigraphic relationship of localities yielding hominids in the Shungura formation, Ethiopia	25
TABLE 3	Excavated sites, in the Koobi Fora region, in relation to stratigraphic markers and selected hominid fossil finds (From *Nature, Lond.*, **262**: 102)	39

LIST OF MAPS

MAP 1	Map of Africa showing hominid sites	xiv
MAP 2	Map of the East African Rift System showing hominid sites	xv

LIST OF FIGURES

FIGURE 1	Plan of areas surveyed east of Lake Turkana	38
FIGURE 2	Plan of the Sterkfontein excavation site. (From *S. Afr. archaeol. Bull.*, **24** : 164)	118
FIGURE 3	Plan of Olduvai Gorge showing hominid sites	167
FIGURE 4	Correlation of magnetic epochs, stratigraphic beds, cultures and hominids at Olduvai Gorge	169

LIST OF PLATES

PLATE 1	Ternifine 1 mandibula. Holotype of *Atlanthropus mauritanicus* Arambourg, 1954. Courtesy of Musée de l'Homme	201
PLATE 2	Yayo 1 fronto-facial fragment. Holotype of *Tchadanthropus uxoris* Coppens, 1965. Courtesy of Musée de l'Homme	202
PLATE 3	Omo 18 corpus mandibulae. Holotype of *Paraustralopithecus aethiopicus* Arambourg & Coppens, 1968. Courtesy of Musée de l'Homme	202
PLATE 4	Fort Ternan FT 46 and 47 maxillae fragments. Holotype of *Kenyapithecus wickeri* Leakey, 1962. Courtesy of P. Andrews	203
PLATE 5	Kanam 1 corpus mandibulae. Holotype of *Homo kanamensis* Leakey, 1935. BM(NH)	203

PLATE 6   Boskop 1 calotte. Holotype of *Homo capensis* Broom, 1917. Courtesy of Port Elizabeth Museum S.A.   204

PLATE 7   Florisbad 1 reconstructed cranium. Holotype of *Homo (Africanthropus) helmei* Dreyer, 1935. Courtesy of Nasionale Museum Bloemfontein .   .   .   205

PLATE 8   Hopefield 1 calvaria. Holotype of *Homo saldanensis* Drennan, 1955. Courtesy of South African Museum   205

PLATE 9   Kromdraai TM 1517 cranium and mandibula. Holotype of *Paranthropus robustus* Broom, 1938. Courtesy of Transvaal Museum, Pretoria   .   .   .   206

PLATE 10   Sterkfontein TM 1511 cranium-palatinum. Holotype of *Australopithecus transvaalensis* Broom, 1936. *Plesianthropus transvaalensis* Broom, 1938. Courtesy of Transvaal Museum, Pretoria   207

PLATE 11   Swartkrans SK6 mandibula. Holotype of *Paranthropus crassidens* Broom, 1949. Courtesy of Transvaal Museum, Pretoria   207

PLATE 12   Makapansgat Limeworks MLD 1 calvaria fragment. Holotype of *Australopithecus prometheus* Dart, 1948. Courtesy of Alun Hughes   .   .   .   208

PLATE 13   Swartkrans SK 15 mandibula. Holotype of *Telanthropus capensis* Broom & Robinson, 1949. Courtesy of Transvaal Museum, Pretoria   .   .   .   208

PLATE 14   Taung 1 skull-dentition. Holotype of *Australopithecus africanus* Dart, 1925. Courtesy of Alun Hughes   209

PLATE 15   Eyasi 1 calotte. Holotype of *Palaeoanthropus njarasensis* Reck & Kohl-Larsen, 1936. Courtesy of Reiner Protsch   210

PLATE 16   Garusi 1 right maxilla. Holotype of *Meganthropus africanus* Weinert, 1950. Courtesy of Reiner Protsch   210

PLATE 17   Olduvai Gorge OH 5 cranium. Holotype of *Zinjanthropus boisei* Leakey, 1959. Courtesy of National Museum, Nairobi   211

PLATE 18   Olduvai Gorge OH 7 mandibula. Part holotype of *Homo habilis* Leakey, Tobias & Napier, 1964. Courtesy of National Museum, Nairobi   .   .   211

PLATE 19   Olduvai Gorge OH 9 calvaria. Holotype of *Homo leakeyi* Heberer, 1963. Courtesy of Alun Hughes   212

PLATE 20   Broken Hill 1 cranium. Holotype of *Homo rhodesiensis* Woodward, 1921. BM (NH)   213

INTRODUCTION

Since the first edition of this part of the Catalogue of Fossil Hominids (1967) became out of print, we have felt it necessary to prepare a new edition incorporating the many new discoveries from all parts of the continent. We feel it is unnecessary to repeat here the full Introduction published in the first edition, but for the convenience of our readers, we have thought it worthwhile to include the key to the layout of each entry. The 18 items of information which we cover are as follows:

1. The place-name designating the fossil. This name usually relates to the site of discovery or a nearby village; we have selected it as the name by which the fossil remains are already known, or shall be known. Where there are a number of hominid fossils known from a single site, discovered and described over a period of years, we have found it advisable to divide them into two or more groups according to their horizon, locus or date of discovery.

2. Description of site and geographical location. Here we have attempted to include a description which would enable a visitor to find the site, and the latitude and longitude to the nearest minute.

    All distances and dimensions are given according to the metric system. Where they were originally given in miles, yards, feet or inches, these have been added in parenthesis.

3. The names of the discoverers and, where applicable, the leader of the expedition responsible for the discovery of the fossils. Where a number of finds were discovered at different times these are listed with their dates of discovery.

4. The geological deposit in which the bones were found, such as river gravel, sand, occupation deposit, cave-earth or breccia. Wherever possible a reference is added for further information. Where references are not included, the contributors have been responsible for the data presented. In this and other items entries in *CHF* * have been quoted where there has been no subsequent publication.

5. Here we indicate whether the bones were a burial. Where there is no definite evidence of burial the line has been left blank. If the find is a burial, an attempt has been made to indicate the layer *from* which the burial was made. In this case, items 6 to 8 refer to the layer of origin of the bones, not the layer into which they were intrusive.

6. The stratigraphical age is stated in general terms (e.g. Pliocene, Upper or Late Pleistocene), and in regional terms. Where opinions differ about this, more than one age is given, followed by appropriate references. Where no reference is given the data represent the editors' opinion based on available publications or personal knowledge. For a full assessment of the stratigraphical age, the reader is advised to consult items 4, 6 and 8. Holocene (Post-Pleistocene) human remains have been included only where they were considered of particular importance.

7. The archaeological context is included here if an industry is present. The reference carries the best description of the industry and not necessarily the current terminology. We have retained as far as possible the term used by the original author quoted, though we have sometimes added the terminology in current use where more applicable.

8. This item indicates the faunal horizon. Whenever possible a selection of animal species associated with each hominid fossil has been included here. The latest revision of the list of associated fauna has been quoted, but in a few cases where no revision has been

**CHF refers to the Catalogue des Hommes Fossiles edited by H. V. Vallois & H. L. Movius (1953).*

published or made known to the editors, the names used in the original publications have been retained.

Where possible the fauna have been listed in a standard order throughout this part of the Catalogue: Mammalia: Primates, Insectivora, Chiroptera, Carnivora, Rodentia, Lagomorpha, Hyracoidea, Tubulidentata, Perissodactyla (e.g. *Equus, Rhinoceros),* Artiodactyla (e.g. *Sus, Hippopotamus, Cervus, Bos),* Proboscidea (e.g. *Elephas),* Sirenia, Cetacea; Aves; Reptilia; Amphibia; Pisces; Invertebrata. Where references are not given, the contributors are responsible for the data presented. References are also included to the best published description of the fauna or flora.

9.   A system of relative dating of bone, antler and dentine has been developed at the Sub-Department of Anthropology, British Museum (Natural History) since 1950, largely through grants-in-aid received from the Wenner-Gren Foundation, under the direction of one of us (K.P.O). This system combines fluorine analysis with uranium estimation by radiometric assay (expressed as equivalent urania, eU308, in parts per million) and nitrogen determination by micro-chemical analysis. It has often proved useful when there has been some doubt as to whether a fossil bone or tooth is contemporaneous with its matrix, derived from an older layer or intrusive by burial from a younger horizon. The results of analysing fossil hominid material have been included in the Catalogue. A few of these results have already been published elsewhere, but the majority were obtained especially for this Catalogue. Unless stated otherwise, the work was done by the British Museum (Natural History) and associated laboratories, mainly the Laboratory of the Government Chemist, London, and the Micro-analytical Laboratory, 10 Carlton Road, Oxford. Wherever possible the analyses of the hominid specimens have been compared with analyses of animal bones from the same or adjacent layers. In the Laboratory of the Government Chemist, fluorine determinations were accompanied by phosphate determinations. If comparison is made between the F/P ratio rather than between the fluorine content of the samples the complicating factor of any contamination by adventitious mineral matter is eliminated. The ratio is usually expressed as $\dfrac{F\%}{P_2O_5\%} \times 100$. For convenience this has been stated in the Catalogue as 100F/P205. Assays for the four elements (F, P, N and U) can be made on a single sample weighing about one gram. Where only 20 mg can be spared F, P and N are usually determinable. Estimations of the fluorine content can be made on as little as 1 mg by the X-ray powder diffraction technique, most usefully applicable where a series of samples is available for comparison.

For further information about relative dating by analytical methods see K. P. Oakley 1963, Fluorine, Uranium and Nitrogen Dating of Bone *in* E. Pyddoke (ed.) *The Scientist and Archaeology,* London: 111–119; K. P. Oakley 1969, Analytical Methods of Dating Bones, *in* D. R. Brothwell & E. S. Higgs (eds.) 1969, *Science in Archaeology,* 2nd Edn, London.

10.   Absolute dates are expressed in years before present, conventionally taken to be 1950, and indicated BP. In this Catalogue, all radiocarbon dates are 'uncalibrated'. (In some publications 'uncalibrated' dates are indicated by bp.) Dates above about 500,000 BP are expressed in millions and designated Myr. Three classes of absolute dating have been distinguished:

A1 dating. Direct determination of the age of the specimen itself from internal evidence; for example, by measuring the carbon-14 radioactivity of residual protein (collagen), or the degree of racemization of amino-acids in human bone.

A2 dating.   Direct determination of the age of the source deposit; e.g. potassium argon (K/Ar) measurements of volcanic minerals where appropriate; thorium/uranium (Th/U) measurements in calcareous rocks; and radiocarbon measurements on the protein collagen, if it has survived in adequate quantity in unburnt bones, dentine or antler, or on the carbon in charred animal remains, charcoal or wood associated with fossil hominids less than 60,000 years old.

A3 dating.   The age of a specimen in years inferred by correlation of the source-bed with a deposit whose age has been determined by potassium/argon, radiocarbon or other chronometric method.

In most instances dates have been quoted without supporting references to published literature, but a sample number has been given wherever possible after initials indicating the laboratory in question.   The following radiocarbon laboratories have contributed results quoted in this part of the catalogue:
C = Chicago; GrN = Groningen; GX = Geochron Lab.   I = Isotopes Inc.; L = Lamont Laboratory; Ly =Lyon; M = University of Michigan; N = Riken (Tokyo); Pta = Pretoria; Sa = Saclay; SR = Salisbury; T = Trondheim; UCLA = University of California, Los Angeles; UW = University of Washington; Y = Yale University.

It is emphasized that by agreement among radiocarbon dating laboratories, published results are based on the Libby half-life of radiocarbon of 5570 ± 30 years.

Evidence for the age of a hominid can be derived from the information given in items 4, 6–10 while it may not be clear from any single item.

11.   Our aim is that every fossil bone and tooth should have a site name and number in addition to the Museum Registration Number.   Where bones and teeth are definitely associated as those of one individual, however, one number is used to refer to them all.   In the case of an entire skeleton, therefore, only one number is used, and this follows traditional practice.   The numbering system does not therefore necessarily give an indication of the number of individuals represented, which often is not known.   In large assemblages of skeletal remains (e.g. Late Palaeolithic and Mesolithic burial sites) the numbers usually represent a minimum of individuals.   Where scattered finds have been reported over a long period of time from a single site, the numbers may well exceed the number of individuals present where the association of different bones as skeletons is not clear.

We have found it best to use arabic numerals to avoid confusion with the roman numerals used so frequently for archaeological numbering.   The arabic numbering system relates to successive discoveries rather than order of antiquity and former roman numbers have been superseded and occasionally modified.   In complex associations such as Koobi Fora, Omo, or Swartkrans the finds have not been re-numbered but have been designated by a numbering system that incorporates their published registration numbers.

The age and sex follow the vernacular designation.   The existence of the *mandibula* is stated if it occurs with the cranium.   A *calvaria* is the *cranium* less the facial bones; a *calotte* is the *calvaria* less the basal bones.   Post-cranial remains have only been listed in detail for the most important remains.   In Upper Pleistocene burials the approximate extent only of the post-cranial remains is indicated.   A perfectly complete skeleton is of course extremely rare, and post-cranial remains from the Middle and Lower Pleistocene are very few.

Latin names for osteological elements have been used according to the publication

Nomina Anatomica, 3rd edition, Amsterdam 1963 (approved by the Seventh International Congress of Anatomists, New York, 1960). For example, the skull-cap is termed the calvaria, not the calvarium. Latin has not, however, been used for descriptive terms, such as upper, lower, left, right, part of, and so on. Our aim has in all cases been to make the description as clear and easily understood as possible in the limited space available.

In a few cases of famous discoveries, fossil hominid finds have a well-known vernacular designation or nickname, which does not conform to our system of numbering, but this has been included. The condition of the bones is roughly indicated by the following abbreviations: i = completely intact; f = fragmentary; ff = very fragmentary. Details of the dentition have only been added for the most important and ancient fossils. Parentheses around teeth indicate they are unerupted.

When they are known the museum registration numbers have been added in brackets to the description of the fossils where there is a large series, in order to facilitate identification.

12. The first published report of the discovery. This reference is often of little palaeontological interest, but has been included as it has been found in many instances to be of some historical value.

13. The first full anatomical description of the remains. This is often a monograph in the case of important finds. Where the author gives an indication of racial or specific status this is added for guidance.

14. The most recent revision of the morphological evaluation of the fossil. This item applies mainly to the older finds, which have been re-studied since their original description was published. We have also put a reference in here, where a second author has disagreed with the conclusions of the author of item 13. In all cases we have added the racial or specific status ascribed to the fossil where this is relevant.

We have not attempted in this Catalogue to assess or judge the validity of any Latin names given to fossil specimens. A review of the nomenclature of the Hominidae has been published by one of us elsewhere (B. G. Campbell 1965, The Nomenclature of the Hominidae, *Occ. Pap. R. anthrop. Inst.,* **22,** London).

15. Reference to the best available illustrated account of the fossil.

16. Any other important and relevant publications beyond those already mentioned. We have sometimes included here general review articles covering the site in question and its placing in relation to others of similar age or locality.

17. The postal address of the repository of the fossil, where this is known, together with the museum registration numbers.

18. The postal address of any institute where moulds for casts are held.

As a further guide, bookmarkers have been prepared and are available with each copy of the Catalogue and these carry a key to the 18 items of information presented for each site listed.

After the original edition of the African part of the Catalogue (published in June 1967) became out of print, steps were taken for the preparation of a new edition. This edition is based on material supplied by contributors to the first edition, which has been revised and updated by the correspondents listed under each country. We wish to thank those correspondents who responded to our request for data on discoveries made since 1966. One of us (T.I.M.) has been especially concerned with the compilation of this revised volume.

There are inevitably gaps in our compilation, though where insufficient information was made available by our correspondents, we have in some instances amplified their contributions.   For any errors and for failure to represent the views of our correspondents, we take responsibility.

It was the generosity of the Board of Directors of the Wenner-Gren Foundation for Anthropological Research which enabled us to undertake the preparation of this Catalogue.   We have felt indebted to them throughout the work, and would like to acknowledge the personal interest taken in its progress by Mrs Lita Osmundsen, Director of Research.

KENNETH PAGE OAKLEY
BERNARD GRANT CAMPBELL
THEYA IVITSKY MOLLESON

British Museum (Natural History),
Cromwell Road, London SW7 5BD

1st September, 1976.

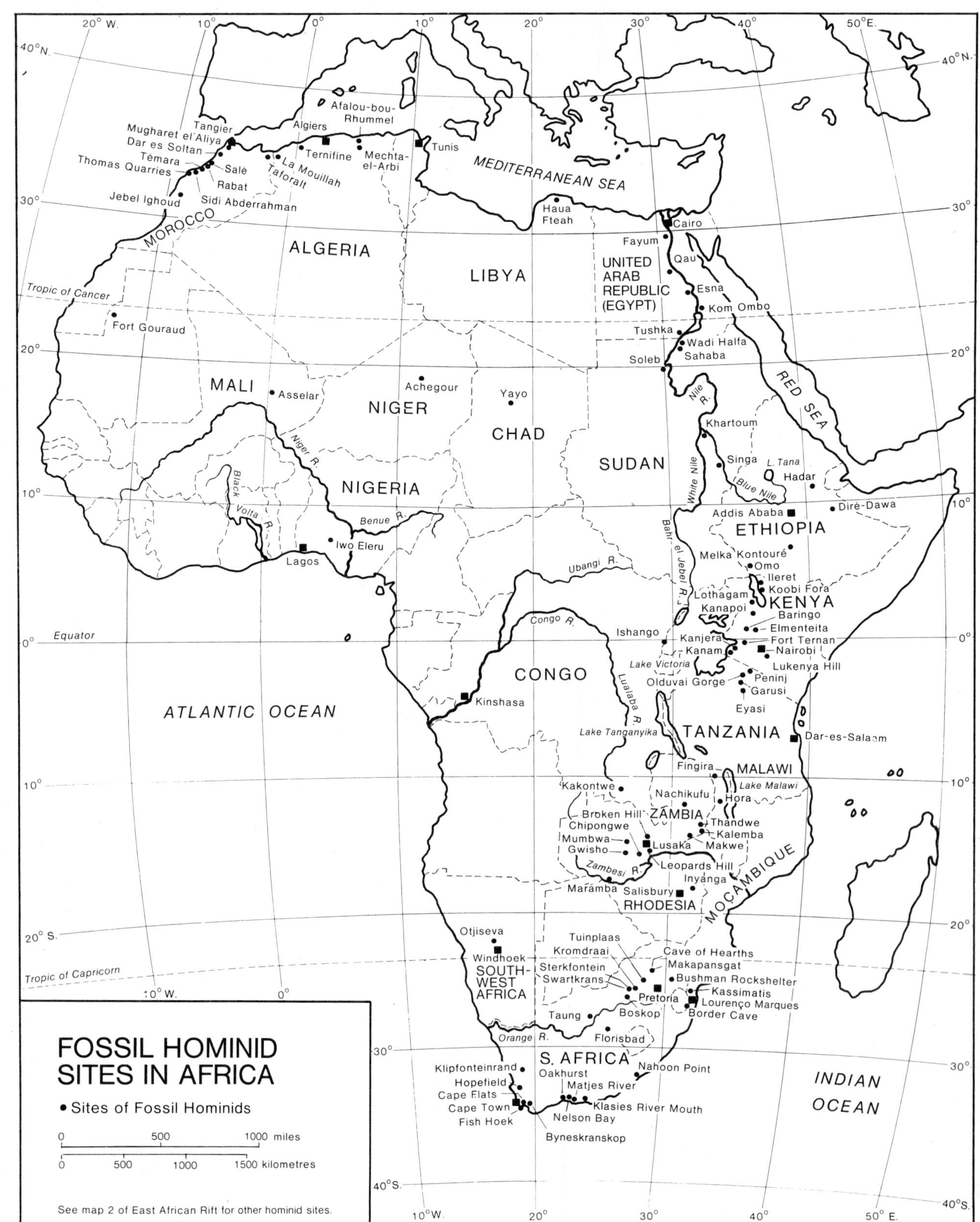

xiv
MAP 1

FOSSIL HOMINID
SITES IN AFRICA
• Sites of Fossil Hominids
0   500   1000 miles
0   500   1000   1500 kilometres
See map 2 of East African Rift for other hominid sites.

MEDITERRANEAN SEA
ATLANTIC OCEAN
INDIAN OCEAN
RED SEA

Equator
Tropic of Cancer
Tropic of Capricorn

40°N.
30°
20°
10°
0°
10°
20°S.
30°
40°S.

20°W.
10°
0°
10°
20°
30°
40°
50°E.

MOROCCO
ALGERIA
LIBYA
UNITED ARAB REPUBLIC (EGYPT)
MALI
NIGER
CHAD
SUDAN
NIGERIA
CONGO
ETHIOPIA
KENYA
TANZANIA
MALAWI
ZAMBIA
MOÇAMBIQUE
RHODESIA
SOUTH-WEST AFRICA
S. AFRICA

Tangier
Mughâret el'Aliya
Dar es Soltan
Témara
Thomas Quarries
Jebel Ighoud
Sidi Abderrahman
Salé
Rabat
Taforalt
La Mouillah
Ternifine
Algiers
Afalou-bou-Rhummel
Mechta-el-Arbi
Tunis
Haua Fteah
Cairo
Fayum
Qau
Esna
Kom Ombo
Tushka
Wadi Halfa
Sahaba
Soleb
Fort Gouraud
Asselar
Achegour
Yayo
Khartoum
Singa
L. Tana
Hadar
Addis Ababa
Diré-Dawa
Melka Kontouré
Omo
Ileret
Koobi Fora
Lothagam
Kanapoi
Baringo
Elmenteita
Ishango
Kanjera
Kanam
Fort Ternan
Nairobi
Lukenya Hill
Olduvai Gorge
Peninj
Garusi
Eyasi
Iwo Eleru
Lagos
Kinshasa
Lake Victoria
Lake Tanganyika
Lake Malawi
Dar-es-Salaam
Fingira
Hora
Kakontwe
Nachikufu
Broken Hill
Chipongwe
Mumbwa
Gwisho
Lusaka
Thandwe
Kalemba
Makwe
Leopards Hill
Inyanga
Marãmba
Salisbury
Otjiseva
Windhoek
Tuinplaas
Kromdraai
Sterkfontein
Swartkrans
Cave of Hearths
Makapansgat
Bushman Rockshelter
Pretoria
Kassimatis
Lourenço Marques
Border Cave
Taung
Boskop
Florisbad
Nahoon Point
Klipfonteinrand
Oakhurst
Matjes River
Hopefield
Cape Flats
Cape Town
Fish Hoek
Klasies River Mouth
Nelson Bay
Byneskranskop

Niger R.
Black Volta R.
Benue R.
Ubangi R.
Congo R.
Lualaba R.
Bahr el Jebel R.
White Nile
Blue Nile
Nile R.
Zambesi R.
Orange R.

33°E.
41°E.
7°N
7°N.
ETHIOPIA
KENYA
TANZANIA
Usno R.
Omo R.
Kibish R.
Usno
Kibish
Shungura
Ileret
Lake Turkana
Koobi Fora
Lothagam
Kanapoi
Lukeino
Ngorora
Chesowanja
Baringo
Loboi
Fort Ternan
Kanjera
Gambles Cave
Kanam
Elmenteita
Lake Victoria
Naivasha
Nairobi
Lukenya Hill
Peninj
Olduvai
Garusi
Laetolil
Eyasi
Arusha
FOSSIL HOMINID SITES
IN EAST AFRICA
Hominid site
Faults
0   50   100 miles
0   50   100   150 kilometres
33°E.
41°E.
4°S
4°S.

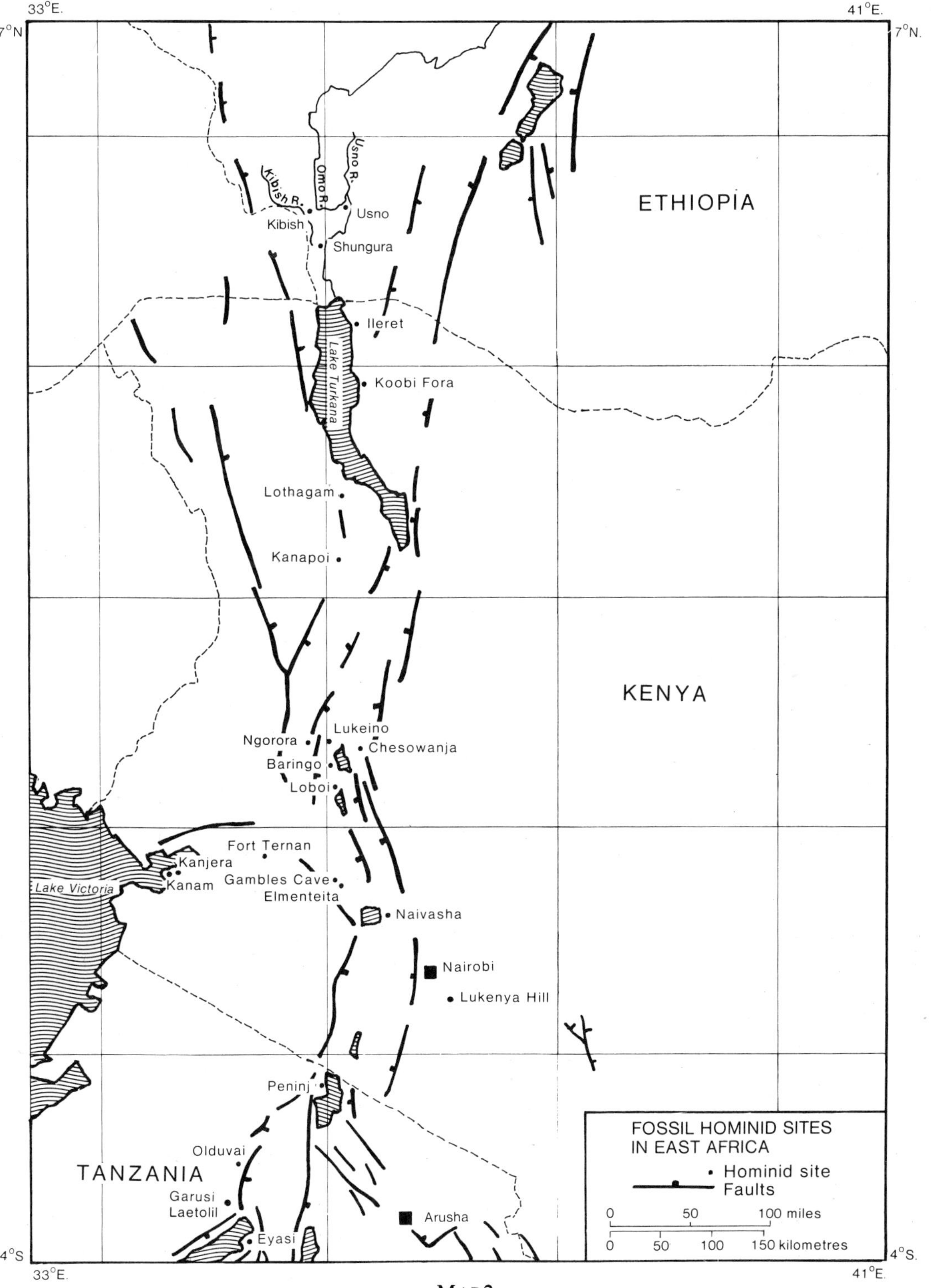

MAP 2

ALGERIA

Lionel BALOUT

Institut de Paléontologie Humaine,
1 rue René Panhard,
75-Paris 13,
France

EDITORS' NOTE

A large number of skeletal remains from Algeria fall near the Pleistocene/Holocene border.   These may be divided into two groups:

      a.   Ibero-Maurusian (= Oranian)

      b.   Capsian

Only a few selected sites from these groups are included in the Catalogue.   Skeletal material has also been reported from the following Oranian sites:

      Alain (rock-shelter)

      Ali Bacha (cave)

      Cap Ténès (escargotière in cave)

      Champlain (escargotière)

      Columnata (rock-shelter)

      Gambetta (escargotière)

      Kef Oum Touiza (rock-shelter)

      La Tranchée (cave)

Two detailed accounts of Columnata have been published: C. Maître 1965, Inventaire des hommes fossiles de Columnata (Tiaret), *Libyca*, **13** : 9–26.   M-C. Chamla, J-N. Birabin and J. Dastugue 1970, Les Hommes Épipaléolithiques de Columnata (Algérie Occidentale), *Mém. Centre Rech. Anthrop. Préhist. Ethnogr.*, **15** : 1–147.

Capsian skeletal material has been omitted with the exception of that from the type site of the Mechta-el-Arbi race, Mechta-el-Arbi itself.   Human remains have also been reported from the following Capsian sites:

      Ain Bahir

      Aioun Beriche

      Bekkaria

      Bir Oum Ali

      Le Chacal

      Jebel Taya

      Khanguet el Mouhaad

      Kilomètre 3,200 near Tebessa

      Oued Medfoun

      Mesloug

      Site 51 near Aïn Beïda

All archaeological human material known up to 1954 from Algeria is listed in L. Balout 1955, *Les Hommes Préhistoriques du Maghred et du Sahara*, Alger.

AFALOU-BOU-RHUMMEL

1. Afalou-bou-Rhummel.
2. Rock-shelter in the cliff overlooking the road from Bougie to Djidjelli, N of Traziboun, 3 km E of the Oued Agrioun.   36° 45′ N, 5° 35′ E.
3. C. Arambourg, April 1928; April–May and October 1929; and autumn 1930.
4. Cave-filling.   Level I: 3·25–4 m grey ash and angular stones.   Level III: 5 m, reddish compact ' bone bed '.   C. Arambourg 1934, *in* C. Arambourg, M. Boule, H. V. Vallois, and R. Verneau 1934, Les grottes paléolithiques des Beni-Segoual (Algérie), *Archs Inst. Paléont. hum.,* **13** : 1–36.
5. Bones in an ossuary (Level I) except Afalou 28 (with 16)—a single burial in Level III below sterile layer.   C. Arambourg 1934: 16–20.
6. Late Pleistocene/Holocene.
7. Ibero-Maurusian (= Oranian).   Microlithic blades with retouched backs of La Mouillah type; polished bone artifacts.   C. Arambourg 1934: 68–81.
8. *Hystrix cristata, Ammotragus dervia, Bos primigenius, Gazella dorcas;* Invertebrata: *Mytilus, Patella, Trochus, Helix.*   C. Arambourg 1934: 26–67.
9. ————
10. ————
11. More than 50 individuals: 26 male, 14 female, 6 juvenile.   In 9 cases crania can be associated with the post-cranial skeleton:
     Afalou   1.   adolescent: skeleton lacking rt manus and ulna; 1 tibia, partly calcined.
     Afalou   2.   adult male: complete skeleton (healed fracture of fibula).
     Afalou   3; 3a.   adult female (i) with infant in arms.
     Afalou 10.   adult male: nearly complete skeleton (healed fracture of rt radius and ulna).
     Afalou 11.   skeleton (f).
     Afalou 13.   skeleton lacking radii, ulnae, ossa mani.
     Afalou 16.   child 2–3 yrs: cranium.
     Afalou 25.   skeleton (f).
     Afalou 28.   adult male (i).
12. C. Arambourg 1929, Découverte d'un ossuaire humain du Paléolithique supérieur en Afrique du Nord, *Anthropologie, Paris,* **39** : 219–221.
13. M. Boule, H. V. Vallois, and R. Verneau 1934, *in* C. Arambourg *et al.* 1934: 83–239.   *Homo sapiens* Mechta-el-Arbi race.
14. H. V. Vallois 1952, Diagrammes sagittaux et mensurations individuelles des Hommes fossiles d'Afalou-bou-Rhummel, *Trav. Lab. Anthrop. Archéol. préhist. Mus. Bardo,* **5** : 1–134.
15. C. Arambourg 1934.
16. D. Ferembach 1960, Les Hommes du Mésolithique d'Afrique du Nord et le problème des Isolats, *Bolm Soc. port. Cienc. nat.,* **23** (2) : 1–16.   M. C. Chamla 1970, Les Hommes Épipaléolithiques de Columnata (Algérie Occidentale), *Mém. Cent. Rech. Anthrop. Préhist. Ethnogr. Alger,* **15** : 1–132.
17. Institut de Paléontologie Humaine, 1 rue René Panhard, 75 Paris 13, France.
18. ————

MECHTA-EL-ARBI

1. Mechta-el-Arbi.
2. Shell-midden, half-way between Sétif and Constantine, near Mechta-Châteaudun station, on the estate of G. Mercier.   36° 15′ N, 5° 55′ E.
3. G. Mercier and A. Dubruge, 1907–1923.   American Mission of Logan Museum, 1926–1927.
4. Shell-midden.
5. Burials constituting an ' ossuary '.
6. Holocene.
7. Upper Capsian. G. Mercier 1908, La station préhistorique de Châteaudun-du-Rummel, *Recl Not. Mém. Soc. archéol. Constantine,* **41** : 172–182.
8. None reported.
9. ————
10. A3 : *c.* 8500 BP on basis of C14 dating of Upper Capsian charcoal, from El Mekta, Tunisia: 8400 $\pm$ 400 BP (L–134).
11. Remains of more than 30 individuals, including the type skulls of the Mechta-el-Arbi race:
    Mechta 3 (1912).   adult male: cranium (i).
    Mechta 4 (1912).   adult female: cranium (f).
12. G. Mercier 1908.
13. H. Lagotala 1924, Étude des ossements humains de Mechta-el-Arbi, *Recl Not. Mém. Soc. archéol. Constantine,* **55** : 145–176.
14. C. Arambourg, M. Boule, H. V. Vallois, and R. Verneau 1934, Les grottes paléolithiques des Beni Segoual (Algérie), *Archs Inst. Paléont. hum.,* **13** : 1–242.   Mechta-el-Arbi race.
15. C. Arambourg *et al.* 1934.
16. L. C. Briggs 1950, On three skulls from Mechta-el-Arbi, Algeria, *Am. J. phys. Anthrop.,* **8** : 305–313.   L. Balout and L. C. Briggs 1951, Mechta-el-Arbi, *Trav. Lab. Anthrop. Archéol. préhist. Mus. Bardo,* **3** : 35–36, 49–50.   M. C. Chamla 1970, Les Hommes Épipaléolithiques de Columnata (Algérie Occidentale), *Mém. Cent. Rech. Anthrop. Préhist. Enthnogr. Alger,* **15** : 1–132.
17. University of Minnesota, Minneapolis 14, Minn., U.S.A.   Mechta 3.   The remainder in the following museums: Musée Mercier, Constantine and Musée d'Ethnographie et de Préhistoire du Bardo, 3 rue Franklin D. Roosevelt, Alger; Collections Société Préhistorique Française, 16 rue St Martin, 75 Paris 4; Institut de Paléontologie Humaine, 1 rue René Panhard, 75 Paris 13, France: Logan Museum of Anthropology, Beloit College, Beloit, Wis., U.S.A.
18. ————

LA MOUILLAH

1. La Mouillah.
2. Rock-shelter, 5 km N of Marnia (Oran), on the left bank of the Oued Mouillah, near the bridge on the road from Marnia to Nemours, approximately 34° 55′ N, 1° 45′ W.

3.   A. Barbin, 1908 and 1910.
4.   Occupation layers in rock-shelter.
5.   Burials: head orientated to west.
6.   Late Pleistocene/Holocene.
7.   Ibero-Maurusian (= Oranian) type site.   A. Barbin 1910, 1912, Fouilles des abris
     préhistoriques de La Mouillah, près Marnia, *Bull. trimest. Soc. Géogr. Archéol. Oran*
     1910, **30** : 77–90; 1912, **32** : 389–402.
8.   *Equus mauritanicus, Bos primigenius.*   A. Barbin 1910, 1912.
9.   —————
10. —————
11. Fragments of 15–16 individuals, including 1 calvaria, 4 calottes.
12. P. Pallary 1909, Note sur un gisement paléolithique de la Province d'Oran, *Bull.*
     archéol. Com. Trav. hist. scient.: 341–342.
13. H. Marchand 1936, Les hommes fossiles de La Mouillah (Oran), *Revue Anthrop.*,
     46 : 239–253.   Mechta-el-Arbi race.
14. M. C. Chamla 1970, Les Hommes Épipaléolithiques de Columnata (Algérie Occident-
     ale), *Mém. Cent. Rech. Anthrop. Préhist. Ethnogr. Alger,* **15** : 1–132.
15. H. Marchand 1936.
16. —————
17. Musée d'Ethnographie et de Préhistoire du Bardo, 3 rue Franklin D. Roosevelt.
18. —————

TERNIFINE

1.   Ternifine.
2.   Sand-pit on outskirts of Palikao village, 20 km E of Mascara, near that part of the plain
     of Eghris called Ternifine.   35° 30′ N, 0° 20′ E.
3.   C. Arambourg and R. Hoffstetter, 9 June 1954 (Ternifine 1); June 1954 (Ternifine 2);
     1955 (Ternifine 3 & 4); 1954 and 1955 (isolated teeth).
4.   Lacustrine sands overlying grey clay lining a cuvette: Ternifine 1 at 1·3 m below the
     water-table.   Ternifine 2 and 4 at 1·4 m below the water-table, Ternifine 3 at 3·7 m
     below the water-table.   Also isolated teeth, some from the sand, some from the
     clay.   C. Arambourg 1963, Le Gisement de Ternifine, pt 2, *L'Atlanthropus*
     mauritanicus, Archs Inst. Paléont. hum., **32** : 37–190, pls A–C, I.
5.   —————
6.   Early Middle Pleistocene.   C. Arambourg and R. Hoffstetter 1963, Le Gisement de
     Ternifine, pt 1, Historique et Géologie, *Archs Inst. Paléont. hum.,* **32** : 9–36.
     Lower Amirian.   P. Biberson 1963, Human Evolution in Morocco in the Framework
     of the Paleoclimatic Variations of the Atlantic Pleistocene, *in* F. C. Howell and F.
     Bourlière (eds) 1963, African Ecology and Human Evolution, *Publs Anthrop. Viking*
     Fund, **36** : 430.
7.   Primitive Acheulian.   L. Balout and J. Tixier 1958, L'Acheulian de Ternifine, *C.r. XV*
     Congr. préhist. Fr. : 214–218.
     Moroccan Acheulian Stage I (Rahmanian).   P. Biberson 1961, Le Cadre
     Paléogéographique de la Préhistoire du Maroc atlantique, *Publs Serv. Antiq. Maroc,*
     16 : 116.

8. Lower Middle Pleistocene fauna with *Machairodus, Afrochoerus, Loxodonta atlantica.*   C. Arambourg and R. Hoffstetter 1963.

9. Ternifine 1: F = 2·1%, 100F/P205 = 5·8, U = 18 ppm, N = 0·3%.
    Loxodonta atlantica: F = 2·3%, 100F/P205 = 7·0, eU308 = 35 ppm, N = 0·7%.
    Equus: F = 1·4%, 100F/P205 = 4·6, eU308 = 24 ppm, N = 0·2%.

10. ——————

11. **Ternifine 1.**   adult: mandibula (f), dentes (2C, 2I missing).   **Holotype** of *Atlanthropus mauritanicus* Arambourg, 1954.   Plate 1.
     Ternifine 2.   young adult: 1t mandibula and symphysis (f), P3–4, M1–3.   (Possibly same individual as Ternifine 4.)
     Ternifine 3.   adult male: mandibula (f) dentes (rt I1, 1t, I2, C, P3 missing).
     Ternifine 4.   young adult about 30 yrs or less: rt parietale.   (Possibly same individual as Ternifine 2).
     Isolated dentes: from maxilla 1C, 2 dm 1, 1 dm 2, 3M (f); lower 2I.

12. C. Arambourg and R. Hoffstetter 1954, Découverte en Afrique du Nord de restes humains du paléolithique inférieur, *C.r. hebd. Séanc. Acad. Sci., Paris,* **239** : 72–74 (Ternifine 1 & 2).   C. Arambourg and R. Hoffstetter 1955, Le gisement de Ternifine: Résultats des fouilles de 1955 et découvertes du nouveaux restes d'Atlanthropus, *C.r. hebd. Séanc. Acad. Sci., Paris,* **241** : 431–433.

13. C. Arambourg 1954, L'hominien fossile de Ternifine (Algérie), *C.r. hebd. Séanc. Acad. Sci., Paris,* **239** : 893–895.   *Atlanthropus mauritanicus.*

14. C. Arambourg 1963.

15. C. Arambourg 1963.

16. C. Arambourg 1955, A recent discovery in human paleontology: *Atlanthropus* of Ternifine (Algeria), *Am. J. phys. Anthrop.,* **13** : 191–201.   P. V. Tobias 1968, Middle and Early Upper Pleistocene Members of the Genus *Homo* in Africa, *in* G. Kurth (ed.) 1968, *Evolution und Hominisation,* 2nd edn, Stuttgart: 179.

17. Muséum National d'Histoire Naturelle, 8 rue de Buffon, 75 Paris 5, France.

18. Muséum National d'Histoire Naturelle, 8 rue de Buffon, 75 Paris 5, France.

CHAD

Yves COPPENS

Musée de l'Homme,
Palais de Chaillot,
75-Paris 16,
France

KORO TORO

See under **YAYO**

YAYO

1. Yayo (Koro Toro).
2. Western end of Angamma cliff, Borkou, 11 km N of the western well of Yago, 160 km NW of Koro Toro.   17° 20′ N, 18° 00′ E.
3. F. A. Coppens, 19 March 1961.   Y. Coppens, director C.N.R.S. excavations.
4. Consolidated fluviatile sands.   Y. Coppens 1962, Découverte d'un Australopithéciné dans le Villafranchien du Tchad, *Colloques int. Cent. natn Rech. scient,* **104** : 455–459.
5. ————
6. Late Lower Pleistocene/Early Middle Pleistocene.   Y. Coppens 1965, L'Hominien du Tchad, *C.r. hebd. Séanc. Acad. Sci., Paris,* **260** : 2869–2871.
7. No industry recorded.   Y. Coppens 1962.
8. *Hippopotamus* sp, *Hippopotamus amphibius, Loxodonta atlantica.*   Y. Coppens 1967, Les Faunes de Vertebrés Quaternaires du Tchad, *in* W. W. Bishop and J. D. Clark (eds.) 1967, *Background to Evolution in Africa,* Chicago: 93–94.
9. Associated material: *Loxodonta atlantica* mandibula: F (X-ray) = $1 \cdot 5$–$2 \cdot 4\%$, eU308 = 49 ppm.   *Hippopotamus:* F (X-ray) = $1 \cdot 5$–$2 \cdot 4\%$, eU308 = 24 ppm.   hominid: F (X-ray) = $0 \cdot 6 \pm 0 \cdot 3\%$, U = 19 ppm.
10. ————
11. **Yayo 1.**   young adult: fronto-facial fragment (f).   **Holotype** of *Tchadanthropus uxoris* Coppens, 1965.   Plate 2.
12. Y. Coppens 1961, Découverte d'un Australopithéciné dans le Villafranchien du Tchad, *C.r. hebd. Séanc. Acad. Sci., Paris,* **252** : 3851–3852.
13. Y. Coppens, in preparation.
14. Y. Coppens 1965.   *Tchadanthropus uxoris.*
15. Y. Coppens 1966, Le Tchadanthropus, *Anthropologie, Paris,* **70** : 5–16.
16. P. V. Tobias 1965, Early Man in East Africa, *Science, N.Y.,* **149** : 22–33.   Y. Coppens 1966, An Early Hominid from Chad, *Current Anthrop.,* **7** : 584–585.
17. Muséum National d'Histoire Naturelle, 8 rue de Buffon, 75 Paris 5, France.
18. ————

EGYPT

Fred WENDORF

Department of Anthropology,
Southern Methodist University,
Dallas, Texas, 75222,
U.S.A.

ESNA

1. Esna.
2. Open camp site near the abandoned Coptic Monastery of Dier el Fakhuri, west bank of the Nile river, 5 km N of Esna.   25° 22′ N, 32° 28′ E.
3. F. Wendorf, February 1967.
4. Ballana formation, dune sand unit.   F. Wendorf, R. Said and R. Schild 1970, Egyptian Prehistory: Some New Concepts, *Science N.Y.*, **169** : 1161–1171.
5. Contemporaneous burials in shallow pit.
6. Upper Pleistocene.   F. Wendorf *et al.* 1970.
7. Late Palaeolithic; Fakhurian industry.
8. *Nesokia indicia, Lepus capensis, Alcelaphus buselaphus, Bos primigenius, Gazella rufifrons.*   Pisces: *Clarias, Barbus.*   Mollusca: *Unio.*
9. Esna 1, pelvis: F = 0·78%, N = 0·04%, long bone: N = 0·02%, mastoid: N = 0·02%.   Esna 2, temporal: N = 0·03%.   Four animal bones: F = 1·66–1·89%, N = 0·02%.
10. A2: 18,020 ± 330 BP (I-3416) on basis of radiocarbon dating of *Unio* shell.
11. **Esna 1.**   adult: pelvis (ff), 2 femora (proximal part).
    Esna 2.   juvenile: calvaria (f), mandibula (f), post-cranial skeleton (ff).
12. F. Wendorf, R. Said and R. Schild 1970 : 29.
13. B. H. Butler 1974, Skeletal remains from a Late Paleolithic site near Esna, Egypt, *in* D. Lubell 1974, The Fakhurian, a Late Paleolithic Industry from Upper Egypt, *Geol. Surv. Egypt,* pap. 58 : 176–183.
14. ————
15. ————
16. F. Wendorf and R. Schild 1976, *Prehistory of the Nile Valley,* New York.
17. Department of Anthropology, Southern Methodist University, Dallas, Texas, 75222, U.S.A.
18. ————

FAYUM

1. Fayum (1926).
2. Open site on old shore-line on NE side of Lake Qarūn a few hundred metres SE of Caton-Thompson's Kom K.   29° 34′ N, 30° 57′ E.
3. K. S. Sandford, 26 March 1926.
4. Fayum Lake Beds at 10 m above sea level.
5. ————
6. Fayum Lake Beds.
7. Fayum Neolithic inferred: concave-base arrow-heads associated with the Lake Beds fauna.   G. Caton-Thompson and E. W. Gardner 1934, *The Desert Fayum,* London, **1** : 72.
8. *Elephas africanus.*   D. M. S. Watson 1934, *in* G. Caton-Thompson and E. W. Gardner 1934, **1** : 72.   Inferred contemporaneous with fauna of Fayum Neolithic Lake Beds.

9. Human calotte: F = 1·2%, 100F/P205 = 3·9, eU308 = 7 ppm, N = 0·1%. *Hippopotamus* vertebra, Fayum Neolithic Lake Beds: F = 1·1%, 100F/P205 = 4·2, eU308 = 5 ppm, N = 0·2%.
10. About 6000 BP (Ed.).
11. **Fayum 1.**   adult female?: calotte (parietalia, occipitale (ff)) with hard sand natural endocast.
12. ——————
13. ——————
14. ——————
15. ——————
16. ——————
17. British Museum (Natural History), Cromwell Road, London SW7 5BD, England.   Reg. No. ZD 1960.1813.
18. ——————

1. Fayum (1937).
2. Open site on N side of Lake Qarūn at about 18 m (50–55 ft) above sea level near Caton-Thompson's Kom W.   Approximately 29° 35′ N, 30° 48′ E.
3. C. Townsend, 1937.
4. Lacustrine hard sandrock (not clay).   G. Caton-Thompson and E. W. Gardner 1934, *The Desert Fayum,* London, 1 : 22.   G. Caton-Thompson 1964, *in lit.*
5. Extended burial in rectilinear grave, possibly Neolithic but Old Kingdom age not impossible.   G. Caton-Thompson 1964, *in lit.*
6. Holocene.
7. Possibly Fayum Neolithic on basis of incisor evulsion and propinquity of flint artifact site.   C. Caton-Thompson 1964, *in lit.*
8. None reported.
9. ——————
10. About 6000 BP (Ed.).
11. **Fayum 2.**   cranium, mandibula.
12. Anon. 1938, *Ill. Lond. News,* 5 February: 212.
13. ——————
14. A. J. Arkell 1953, *CHF.*: 220.   Negroid of Old Kingdom type.
15. Anon. 1938.
16. ——————
17. Department of Anatomy, Cairo University, Cairo.
18. ——————

1. Fayum (1962).
2. Open site approximately 2·8 km SW of Qasr es Sagha Temple 50 – 100 m N of edge of Site R and about 140 m S of Dynastic Cemetery at Qasr es Sagha.

3. E. L. Simons, December 1962.   Director, Yale University Expedition to the Fayum.
4. Lake beds.   G. Caton-Thompson 1934, *The Desert Fayum,* London, **1** : 14–15.
5. ————
6. Holocene.
7. Microlithic.   E. L. Simons 1964, *in lit.*
8. Antelope, *Hippopotamus.*   E. L. Simons 1964, *in lit.*
9. Human parietal: $F = 1\cdot5\%$, $100F/P_2O_5 = 4\cdot7$, $eU_3O_8 = 5$ ppm, $N = 0\cdot2\%$.   Animal bone from same level: $F = 1\cdot4\%$, $100F/P_2O_5 = 4\cdot5$, $eU_3O_8 = 9$ ppm, $N = 0\cdot1\%$.
10. Presumed same age as Fayum 1.   (Ed.).
11. **Fayum 3.**   lt parietale (ff).
12. ————
13. ————
14. ————
15. ————
16. ————
17. Peabody Museum of Natural History, Yale University, New Haven, Conn., U.S.A.   Collections in Vertebrate Paleontology 20903.
18. ————

KOM OMBO

1. Kom Ombo (1926).
2. 1·6 km N of the Jebel Silsila station, and about 0·4 km E of the railway line, Kom Ombo plain.   24° 20′ N, 32° 50′ E.
3. K. S. Sandford, 22 March 1926.
4. Silts in Sebilian Gravels.   K. S. Sandford 1934, *Palaeolithic Man and the Nile Valley in Upper and Middle Egypt,* Chicago : 85–86.
5. ————
6. Upper Pleistocene: Sebilian Gravels.   K. S. Sandford 1934: 83–87.
7. Lower or Middle Sebilian.   K. S. Sandford 1934.
8. *Equus asinus, Bos primigenius.*   Aves: *Struthio.*   K. S. Sandford 1934: 86.
9. Human bone in same state of mineralization as fauna.   K. S. Sandford 1934 : 86.
10. A3: *c.* 13,000 BP on basis of dating of Kom Ombo 2.
11. **Kom Ombo 1.**   calvaria.
12. K. S. Sandford 1934 : 86.
13. ————
14. K. S. Sandford (quoting G. Elliot Smith) 1934 : 86.   Cf. Pre-dynastic Egyptian.
15. ————
16. ————
17. Whereabouts unknown since 1939: possibly lost during 1939–1945 war.
18. ————

1. Kom Ombo (1963).
2. Open site, Jebel Silsila 2B, Sebilian area, E side of Nile River, about 4 km N of Jebel Silsila Station, 1 km E of the Cairo-Aswan railway line, Kom Ombo Plain, Upper Egypt.   Site later destroyed by agriculture.   24° 20′ N, 32° 50′ E.
3. M. A. Baumhoff, 4 February 1963.   Yale University Prehistoric Expedition to Nubia.
4. Younger Channel Silts (Jebel Silsila formation, Daurau Member).   Derived in fluvial beds, right bank of former branch of Nile, 96 m above sea-level, 7·5 m above present floodplain.   C. A. Reed 1965, A Human Frontal Bone from the Late Pleistocene of the Kom Ombo Plain, Upper Egypt, *Man,* **65** : 101–104.   K. W. Butzer and C. L. Hansen 1965, On Pleistocene Evolution of the Nile Valley in Southern Egypt, *Canad. Geog.,* **9** : 73–84.   K. W. Butzer and C. L. Hansen 1968, Desert and river in Nubia: Geomorphology and prehistoric environments at the Aswan Reservoir, *U. Wisconsin Press,* Madison: 170–177.
5. ─────────
6. Uppermost Pleistocene.   C. A. Reed 1965.
7. Middle Sebilian.   E. Vignard 1923, Une Nouvelle industrie lithique, le ' Sebilian', *Bull. Inst. fr. Arch. orient.,* **22** : 1–76.   C. A. Reed 1965.   C. A. Reed, M. A. Baumhoff, K. W. Butzer, H. Walter and D. S. Boloyan 1967, Preliminary report on the Archaeological Aspects of the research of the Yale University Prehistoric Expedition to Nubia, 1962–1963, Department of Antiquities, Government of the United Arab Republic : 145–156.
8. *Alcelaphus, Bos primigenius, Equus asinus, Hippopotamus amphibius, Ammotragus lervia, Gazella dorcas;* Pisces: *Lates niloticus, Clarias.*   C. S. Churcher 1972, Late Pleistocene vertebrates from archaeological sites in the plain of Kom Ombo, Upper Egypt, *Life Sci. Contr. r. Ont. Mus.,* **82**: 1–172.
9. Human frontal: $F = 1\cdot2\%$, $100F/P205 = 4\cdot3$, $eU308 = 36$ ppm, $N = 0\cdot2\%$.   *Hippopotamus* bone from Sebilian silts: $F = 1\cdot0\%$, $100F/P205 = 4\cdot1$, $eU308 = 6$ ppm, $N = $ nil.   K. P. Oakley 1965, The Antiquity of the New Kom Ombo Skull, *Man,* **65** : 104.
10. A2: $13,070 \pm 160$ years BP (Y-1375) on basis of C14 dating of charcoal; $13,560 \pm 120$ years BP (Y-1447), on basis of C14 dating of *Unio* shells.   C. A. Reed 1965.
11. **Kom Ombo 2.**   adult male: frontale (f) (No. 20333).   Possibly also parietale (f) (No. 21169), rt parietale (f) (No. 21174).
     Other human remains: occipitale (ff) (No. 21170), lt parietale (ff) (No. 21171).   2 parietal fragments (No. 21172), frontale (f) (No. 21173), bone fragment (No. 21175).
12. C. A. Reed 1965.
13. C. A. Reed 1965.
14. ─────────
15. C. A. Reed 1965.
16. P. E. L. Smith 1967, New investigations in the late Pleistocene archaeology of the Kom Ombo Plain (Upper Egypt), *Quaternaria,* **9** : 141–152.   P. E. L. Smith 1976, Stone-Age Man on the Nile, *Sci. Am.,* **235**: 30–38.
17. Peabody Museum of Natural History, Yale University, New Haven, Conn., U.S.A.   Collections in Vertebrate Paleontology, 2033, 21169–21175.
18. ─────────

QAU

1. Qau (Kau).
2. Two grave shafts filled during Dynastic times with fossil bones from unlocalized surface outcrops. Cemeteries near Qau-el-Kabir, on right bank of Nile about 50 km S of Asyut. 26° 57′ N, 31° 33′ E.
3. G. Brunton, January 1923. W. M. Flinders Petrie, 1924.
4. Derivation from formerly exposed gravels of the Nile. K. S. Sandford 1929, The Pliocene and Pleistocene Deposits of Wadi Qena and of the Nile Valley between Luxor and Assiut: VIII. The Fossil Bones found at Qau, and Beds proved in Borings, *Q. Jl geol. Soc. Lond.*, **85** : 536–541.
5. Shaft fillings with large quantities of fossilized bones and *Hippopotamus* ivory, all presumably obtained from the Sebilian gravels on the Kom Ombo plain. K. S. Sandford 1934, *Palaeolithic Man and the Nile Valley in Upper and Middle Egypt,* Chicago: 85–86. *CHF*: 219.
6. Upper Pleistocene, Sebilian gravels? K. S. Sandford 1934: 83–87.
7. No originally associated artifacts recorded. K. S. Sandford 1934: 85.
8. *Gazella isabella, Sus.* Anon 1923, Fossil Human Bones, possibly of Pleistocene age, found in Egypt, *Nature, Lond.,* **112** : 250. *Crocodylus* sp. nov. D. M. S. Watson *in* K. S. Sandford 1934: 86. *Homoioceras?* A. J. Arkell 1953, *CHF*: 219.
9. Human femur (light): $F = 1\cdot2\%$, $100F/P205 = 4\cdot1$, $eU308 = 15$ ppm, $N = 0\cdot1\%$.
   Fossil ungulate bone: $F = 1\cdot0\%$, $100F/P205 = 4\cdot1$, $eU308 = 3$ ppm, $N = $ nil.
   Human tibia (dark): $F = 2\cdot3\%$, $100F/P205 = 7\cdot3$, $eU308 = 27$ ppm, $N = $ nil.
   Fossil ungulate bone: $F = 1\cdot3$, $100F/P205 = 3\cdot5$, $eU308 = 4$ ppm, $N = $ nil.
10. ———————
11. **Qau 1–10.** fragments include 3 frontalia (f), 4 parietalia (f), 2 occipitalia (ff), rt temporale, 2 rt rami mandibulae, 2 lower Il, axis, 2 rt humeri (f), 3 ilia (ff), fragments of femora (3 heads) (ff), fragments of tibiae (mid-shaft) (ff), 1 lt patella.
12. Anon. 1923. W. M. Flinders Petrie 1925, Early Man in Egypt, *Man,* **25** : 130.
13. D. E. Derry 1923, ms. Report on Fossil Human Bones from Qau, copies filed in Department of Egyptology, University College, Gower Street, London WC1, and in British Museum (Natural History), Cromwell Road, London SW7 5BD.
14. K. S. Sandford (quoting A. Keith) 1934: 86. Cf. Pre-dynastic Egyptian.
15. ———————
16. W. M. Flinders Petrie 1925. G. Brunton 1927, 1930, *Qau and Badari,* London, **1** : 1–2, 12; **3** : 15, 18, 20.
17. Cranial fragments: whereabouts unknown, possibly lost during 1939–1945 war. Femur and tibia fragments: British Museum (Natural History), Cromwell Road, London SW7 5BD, England. Reg. No. EM 676–687.
18. ———————

TUSHKA

1. Tushka.
2. Open camp site 8905 with associated graveyard, west bank of the Nile River, 30 km N of the temples of Abu Simbel.   22° 30′ N, 31° 45′ E.
3. J. Hester and P. Hoebler, under direction of F. Wendorf, 1965.
4. Ballana formation, dune sand unit.   C. C. Albritton 1968, *in* F. Wendorf (ed.) 1968, *The Prehistory of Nubia,* Dallas : 856–864.   J. de Heinzelin 1968, *in* F. Wendorf (ed.) 1968 : 45.
5. Burials within Ballana formation.
6. Upper Pleistocene or end Pleistocene.
7. Late Palaeolithic: Qadan.   F. Wendorf 1968 : 864–946.
8. A. Gautier 1968, *in* F. Wendorf (ed.) 1968 : 80–99.
9. Human long-bone, level 6: eU308 = 13 ppm, N = 0·2%; level 14: eU308 = 10 ppm, N = 0·1%.   Animal bone, level 10(a): eU308 = 8 ppm, N = 0·1%; level 7(a): eU308 = 10 ppm, N = nil.
10. A2: 12,000–10,000 BP.   F. Wendorf 1968 : 940.
11. **Tushka 1–19.**   fragmentary remains of one infant and 18 adults.
12. F. Wendorf 1968 : 869–874.
13. J. E. Anderson 1968, *in* F. Wendorf (ed.) 1968 : 1016–1018.   Cf. Mechta-el-Arbi.
14. ————
15. J. E. Anderson 1968.
16. F. Wendorf, R. Said and R. Schild 1970, Egyptian Prehistory: Some New Concepts, *Science, N.Y.,* **169** : 1161–1171.
17. Egyptian Museum, Tahir Square, Cairo (2 individuals).   Department of Anthropology, Southern Methodist University, Dallas, Texas, 75222, U.S.A.   (17 individuals).
18. ————

ETHIOPIA

Yves COPPENS

Musée de l'Homme,
Palais de Chaillot,
75-Paris 16,
France

LAKE BESAKA   8° 52′ N, 39° 51′ E.

In 1974, J. D. Clark excavated 6 burials from an ashy, sandy occupation midden at a site adjacent to a low scarp, 1·4 km from the W shore of Lake Besaka, E of Addis Ababa. They are associated with Phase B of the Ethiopian Late Stone Age industry. The estimated age is 5000–7000 BP.

J. D. Clark and M. A. J. Williams 1977, Recent Archaeological Research in South-eastern Ethiopia, *Annls Ethiopie*, in press.

The skeletons are preserved at the National Museum, Addis Ababa.

DIRÉ-DAWA

1. Diré-Dawa.
2. Cave, 'Porc-Épic', 200 m above the mountain-stream which follows the road from Diré-Dawa to Balla peak, 2 km upstream from Diré-Dawa, Harar.   9° 40′ N, 41° 50′ E.
3. H. Breuil, February–March 1933.
4. Within concreted phosphatized layer on rocky ledge to right of cave entrance.   H. Breuil, P. Teilhard de Chardin and P. Wernert 1951, Le Paléolithique du Harrar, *Anthropologie, Paris,* **55** : 226.   Brecciated cave earth underlying thick stalagmite layer.   J. D. Clark 1976, *in lit.*
5. ————
6. Early Upper Pleistocene.   J. D. Clark 1976.
7. Evolved 'Middle Stone Age'.   Breuil *et al.* 1951.   Middle Palaeolithic/Middle Stone Age occupation material.   J. D. Clark 1976.
8. No fauna in basal layer.   Recent fauna in upper ashy layer.   H. Breuil *et al.* 1951 : 226.   Associated fauna comprises antelopes, pig, horse and rodents.   No extinct forms.   J. D. Clark 1976.
9. ————
10. ————
11. **Diré-Dawa 1.**   adult: rt corpus mandibulae, P3–M3, all damaged.
12. H. V. Vallois 1951, La mandibule humaine fossile de la grotte du Porc-Épic, près Diré-Daoua (Abyssinie), *Anthropologie, Paris,* **55** : 231–238.
13. H. V. Vallois 1951.   Neandertaloid.
14. ————
15. H. V. Vallois 1951.
16. P. V. Tobias 1968, Middle and Early Upper Pleistocene Members of the Genus *Homo* in Africa, *in* G. Kurth (ed.) 1968, *Evolution und Hominisation* 2nd edn., Stuttgart: 184.
17. Institut de Paléontologie Humaine, 1 rue René Panhard, 75 Paris 13, France.   Reg. No. 50–2.
18. ————

HADAR

1. Hadar,
2. Surface finds of fossiliferous sediments at Hadar River, tributary of Lower Awash River, 30 km S of Weranzo between Dessie and Assab in the Afar Depression.   11° 00′ N, 40° 30′ E.   (Map 2) p. xv.
3. D. C. Johanson, 30 October 1973 (AL 128–1, AL 129–1a, 1b, 1c); 11 December 1973 (AL 166–9).   A. Asfaw, 16 October 1974 (AL 188–1). M. Kasa, 18 October 1974 (AL 198–1); 5 December 1974 (AL 198–17a, 17b, AL 198–18).   A. Asfaw, 17 October 1974 (AL 199–1, AL 200–1a, 1b).   R. Ciochon, 20 October 1974 (AL 211–1).   G. Corvinus, 27 October 1974 (AL 228–1).   D. Lost, 22 December 1974 (AL 241–14).   A. Asfaw, 16 November (AL 266–1).   A. Dato, 19 November 1974 (AL 277–1).   D. C. Johanson and B. T. Gray, 24 November 1974 (AL 288–1).
4. Hadar Formation: lacustrine sands and muds with volcanic marker beds and hominid fossils at 20–90 m above the Awash River.   (See Table I).   M. Taieb, D. C. Johanson, Y. Coppens, R. Bonnefille and J. Kalb 1974, Découverte d'Hominidés dans les Séries Plio-Pléistocènes d'Hadar (Bassin de l'Awash; Afar, Éthiopie), *C.r. hebd. Séanc. Acad. Sci., Paris,* **279D**: 735–738.   M. Taieb 1974, Évolution Quaternaire du Bassin de l'Awash (Rift Éthiopien et Afar), *Thèse Doctorat d'État,* Faculté des Sciences de Paris.   M. Taieb, D. C. Johanson, Y. Coppens and J. L. Aronson 1976, Geological and

TABLE I

Stratigraphic relationship of hominids from the Hadar Formation

Hadar Formation		*Marker Beds*	*K/Ar Myr*	*Faunal zone*	*Hominids*	*Palaeo-magnetic sequence*
Kada Hadar	KH upper			Upper		
	KH3					N
	KH2					N
	KH1				288	N
		KHT Tuff				
Denen Dora	DD3				188	
	DD2			Middle	241	
	DD1	TT				
Sidi Hakoma	SH upper	Kadada Moumou basalt	3·0		211	R=Mammoth
					266	or Kaena
					277	
	SH3					
	SH2			Lower	228	
					128, 129, 166	
	SH1				198, 199, 200	N
		SHT Tuff	3·1			
			4·1			
			5·3			
Basal Member BM						

Palaeontological background of Hadar hominid site, Afar, Ethiopia, *Nature, Lond.,* **260** : 289–293.

5. ————

6. Upper Pliocene. M. Taieb *et al.* 1974.

7. ————

8. Fauna equivalent to Usno and the lower portion of the Shungura Formation (lower Omo basin), Members A to B: *Loxodonta adaurora, Elephas* cf. *ekorensis, E. recki, Tragelaphus* cf. *nakuae, Notochoerus euilus, Nyanzachoerus pattersoni,* Hexaprotodont Hippopotami, Rhinocerotidae, *Hipparion.* D. C. Johanson, M. Taieb and Y. Coppens 1977, Plio-Pleistocene Hominid Discoveries in Hadar, central Afar, Ethiopia, *in* C. Jolly (ed.) 1977, *Early Hominids of Africa.* M. Taieb *et al.* 1976. F. C. Howell and G. Petter 1976, Indications d'âge données par les carnivores de la formation d'Hadar (Éthiopie orientale), *C.r. hebd. Séanc. Acad. Sci., Paris,* **282D** : 2063–2066.

9. ————

10. A3: 3·0 ± 0·2 Myr on the basis of K/Ar dating of the Sidi Hakoma Tuff 10 m above base of section and of Kadada Moumon basalt 50–60 m above the base. M. Taieb *et al.* 1976.

11. **Hadar AL 128–1**   lt proximal femur (f) (probably associated with A.L. 129).
    Hadar AL 129–1a.   rt distal femur (f)
    Hadar AL 129–1b.   rt proximal tibia (f) (probably associated with A. L. 128).
    Hadar AL 129–1c.   rt proximal femur (f)
    Hadar AL 166–9.   pars petrosa lt temporale (ff).
    Hadar AL 188–1.   rt corpus mandibulae, M2–M3 (ff).
    Hadar AL 198–1.   juvenile lt corpus mandibulae, C, P3, P4, dm 2, M1, M2.
    Hadar AL 198–17a. upper lt I1.
    Hadar AL 198–17b. upper lt I2.
    Hadar AL 198–18.   lower rt I2.
    Hadar AL 199–1.   rt maxilla, C–M3 (f).
    Hadar AL 200–1a,b maxilla, 16 dentes (i), lower rt M1 crown.
    Hadar AL 211–1.   rt proximal femur (f).
    Hadar AL 228–1.   rt distal femur diaphysis (ff).
    Hadar AL 241–14.   lower lt M.
    Hadar AL 266–1.   corpus mandibulae, lt P3–M1 (f), rt P3–M3 (f).
    Hadar AL 277–1.   lt corpus mandibulae, C–M2 (ff).
    Hadar AL 288–1.   female, ' Lucy ': cranial fragments, corpus mandibulae, lt P4, M3, rt P3–M3, sockets of remaining teeth, vertebrae lumbalis (i–ff), costae, scapula (ff), lt os coxae (i), sacrum (i), 2 humeri (i, f), 2 radii (f, ff), 2 ulnae (i, ff), 2 phalanges manus, capitatum, lt femur (i), rt tibia (f), rt distal fibula (ff), rt talus (i).
    Hadar AL 333–1–7. adults and children: cranial, maxillary and mandibular fragments radius, ulna, metacarpalia, phalanges.

12. M. Taieb *et al.* 1974.

13. D. C. Johanson and M. Taieb 1976, Plio-Pleistocene hominid discoveries in Hadar Ethiopia, *Nature, Lond.,* **260** : 293–297. M. Taieb, Y. Coppens, D. C. Johanson and R. Bonnefille 1975, Hominidés de L'Afar Central, Ethiopie (Site d'Hadar, Campagne 1973), *Bull. Mém. Soc. Anthrop.,* **2** : 117–124. (A. L. 128–1, A. L. 129–1a, b, c) cf. *Australopithecus africanus;* (A. L. 166–9) cf. *Australopithecus robustus.* M. Taieb, D.

C. Johanson and Y. Coppens 1975, Expédition internationale de l'Afar, Ethiopie (3e campagne 1974); découverte d'Hominidés plio-pléistocènes à Hadar, *C.r. hedb. Séanc. Acad. Sci. Paris,* **281D** : 1297–1300 (A. L. 166–9, A. L. 211–1, A. L. 188–1) cf. *A. robustus*: (A. L. 128–1, A. L. 129–1a–c, A. L. 288–1) cf. *A. africanus* (A. L. 199–1, A. L. 200–1, A. L. 266–1). Cf. *Homo.*   D. C. Johanson and Y. Coppens 1976, A Preliminary Anatomical Diagnosis of the First Plio/Pleistocene Hominid Discoveries in the Central Afar, Ethiopia, *Am. J. phys. Anthrop.,* **45** : 217–234 (AL 129, a, b, c, 166–9).

14. ————————
15. M. Taieb, D. C. Johanson and Y. Coppens 1975, Expédition internationale de l'Afar, Ethiopie (3e campagne 1974): découverte d'Hominidés pliopléistocènes à Hadar, *C.r. hebd. Séanc. Acad. Sci., Paris,* **281D** : 1297–1300.   D. C. Johanson *et al.* 1976 (AL 200–1a, AL 199–1, AL 288–1).   D. C. Johanson and Y. Coppens 1976.
16. ————————
17. National Museum, P.O. Box 5648, Addis Ababa.
18. ————————

MELKA KONTOURÉ

1. Gomboré II.
2. Open site Gomboré II on right bank of Awash River at Melka-Kontouré, 50 km S of Addis Ababa.   8° 42′ N, 38° 37′ E.
3. C. Brahimi, 1973.
4. Occupation soil on upper Fluviatile deposits.   J. Chavaillon, C. Brahimi and Y. Coppens 1974, Première découverte d'Hominidé dans l'un des sites acheuléens de Melka-Kunturé (Ethiopie), *C.r. hebd. Séanc. Acad. Sci., Paris,* **278 D** : 3299–3302.
5. ————————
6. Gomboré II level of Middle Pleistocene deposits.
7. Acheulian.   J. Chavaillon 1973, Chronologie des niveaux paléolithoques de Melka-Kunture (Ethiopie), *C.r. hebd. Séanc. Acad. Sci., Paris,* **276 D** : 1533–1536.
8. Fauna includes Rodentia; *Equus, Hippopotamus;* Ungulata.   J. Chavaillon *et al.* 1974.
9. ————————
10. ————————
11. **Melka-Kontouré 1.**   lt parietale (f).
12. J. Chavaillon *et al.* 1974.
13. J. Chavaillon *et al.* 1974.   Cf. *Homo erectus.*
14. J. Chavaillon and Y. Coppens 1975, Découverte d'Hominidé dans un site Acheuléen de Melka-Kunturé (Ethiopie), *Bull. Mém. Soc. Anthrop. Paris,* **2** : 125–128.
15. J. Chavaillon *et al.* 1974.
16. ————————
17. Institute Ethiopien d'Archéologie, P.O. Box 5648, Addis Ababa.
18. ————————

OMO

1. Kibish (KHS).
2. Open site, 1·8 km NW of Kenya Camp near Shiangoro Village, E of Omo river.   5° 24′ N, 35° 57′ E.
3. Expedition under the direction of R. E. F. Leakey, 1967.
4. From disconformity in fluviatile deposits of silty clay loam in adjoining delta.   Member I e/d, Kibish Formation.   K. W. Butzer 1969, Geological Interpretation of Two Pleistocene Hominid Sites in the Lower Omo Basin, *Nature, Lond.*, **222** : 1138–1140.
5. ————
6. Late Middle to Early Upper Pleistocene.   K. W. Butzer 1969.
7. Surface finds of a few stone artifacts; flake debris *in situ*.   R. E. F. Leakey 1969, Early *Homo sapiens* Remains from the Omo River Region of South-west Ethiopia, *Nature, Lond.*, **222** : 1137–1138.
8. *Colobus.*   Complete skeleton of *Synerus* aff. *caffer, Elephas loxodonta* (primitive variety), *E. recki.*   R. E. F. Leakey 1969.
9. ————
10. A3 : 39,900 BP on basis of C14 dating of *Etheria* shells in overlying deposit, member If; *c.* 130,000 BP (L–1303–J) on basis of Th230/U234 dating of the same shells. K. W. Butzer, F. H. Brown and D. L. Thurber 1969, Horizontal Sediments of the Lower Omo Valley: Kibish Formation, *Quaternaria,* **11** : 15–29.
11. **Omo 1.**   adult: calvaria (f), maxillae, zygoma, mandibula (f), 2 dentes, 3 vertebrae cervicalis, 8 vertebrae thoracicae (f), 1 vertebra lumbalis (f), costae; 1t clavicula (i), rt calvicula (f), processi coracoidei, 2 humeri (f), 2 radii (f), rt ulna (f); 2 carpi, 4 metacarpalia (f), phalanges of rt manus; rt femur (f), 2 tibiae (f), rt fibula (f), tarsi, metatarsalia and phalanges of rt pes.
12. R. E. F. Leakey 1969.
13. M. Day 1969, Omo Human Skeletal Remains, *Nature, Lond.*, **222** : 1140–1143. *Homo sapiens* cf. Broken Hill.
14. ————
15. M. Day 1969.
16. G. P. Rightmire 1976, Relationships of Middle and Upper Pleistocene hominids from sub-Saharan Africa, *Nature, Lond.,* **260** : 238–240.
17. National Museum, Addis Ababa.
18. ————

1. Kibish (PHS).
2. Open site, 4·8 km NW of Kenya Camp, W of Omo River.   5° 25′ N, 35° 55′ E.
3. Expedition under the direction of R. E. F. Leakey, 1967.
4. Surface find on fluviatile clay, eroded from upper part of member I unit 5–6 Kibish formation (equivalent to level of Omo 1).   K. W. Butzer 1969, Geological Interpretation of Two Pleistocene Hominid Sites in the Lower Omo Basin, *Nature, Lond.*, **222** : 1138–1140.
5. ————
6. Late Middle to Early Upper Pleistocene.   K. W. Butzer 1969.

7.   None reported.

8.   None reported.

9.   ————

10.   A3: >39,900 BP on basis of C14 dating of *Etheria* shells in overlying deposit, Member If; *c.* 130,000 BP (L–1303–J) on basis of Th230/U234 dating of the same shells.   K. W. Butzer, F. H. Brown and D. L. Thurber 1969, Horizontal Sediments of the Lower Omo Valley: Kibish Formation, *Quaternaria,* **11** : 15–29.

11.   **Omo 2.***   adult: calvaria.

12.   R. E. F. Leakey 1969, Early *Homo sapiens* Remains from the Omo river Region of South-west Ethiopia, *Nature, Lond.,* **222** : 1137–1138.

13.   M. Day 1969, Omo Human Skeletal Remains, *Nature, Lond.,* **222** : 1140–1143. *Homo sapiens* cf. Swanscombe and Skhūl.

14.   ————

15.   M. Day 1969.

16.   ————

17.   National Museum, Addis Ababa.

18.   ————

---

*Omo 3, a left fronto-parietal fragment was also found at this time in the Kibish formation deposits, but details of its provenance are not published.   It is described in M. Day 1969.

1. Shungura.
2. Series of 79 localities covering about 200 square km on W bank of Omo river.   4° 35′ to 5° 10′ N, 36° 00′ E.   (Map 2) p. xv.
3. Omo Research Expedition under direction of C. Arambourg, F. Clark Howell and Y. Coppens, 1967–72.
4. Omo Beds = Shungura formation: sedimentary deposits interbedded with volcanic tuffs designated A to J (from older to younger.   See Table 2).   F. H. Brown 1969, Observations on the stratigraphy and radiometric age of the 'Omo Beds', lower Omo basin, southern Ethiopia, *Quaternaria,* 11 : 7–14.   J. de Heinzelin, F. H. Brown and F. C. Howell 1970, Pliocene/Pleistocene formations in the lower Omo basin, southern Ethiopia, *Quaternaria,* 13 : 247–268.   R. Bonnefille, F. H. Brown, J. Chavaillon, Y. Coppens, P. Haesaerts, J. de Heinzelin and F. C. Howell 1973, Situation stratigraphique des localités à Hominidés des gisements plio-pléistocènes de l'Omo en Ethiopie *C.r. hebd. Séanc. Acad. Sci. Paris,* **276D** : 2781–2784, 2879–2882.
5. ────────
6. Upper Pliocene to Lower Pleistocene.   F. C. Howell and Y. Coppens 1974, Inventory of remains of Hominidae from Pliocene/Pleistocene Formations of the Lower Omo Basin, Ethiopia (1967–1972), *Am. J. phys. Anthrop.,* **40** : 1–16.
7. Member E: Oldowan chopper.   J. Chavaillon 1970, Découverte d'un niveau oldowayen dans la basse vallée de l'Omo (Ethiopie), *C.r. Soc. préhist. franc.,* **67** : 7–11.
   Member F: small shattered quartz lumps, flakes and pebbles.   H. V. Merrick, J. de Heinzelin, P. Haesaerts and F. C. Howell 1973, Archaeological Occurrences of Early Pleistocene Age from the Shungura Formation, Lower Omo Valley, Ethiopia, *Nature, Lond.,* **242** : 572–575.   Y. Coppens, J. Chavaillon and M. Bedan 1973, Résultats de la nouvelle mission de l'Omo (campagne 1972)—Découverte de restes d'Hominidés et d'une industrie sur éclats, *C.r. hebd. Séanc. Acad. Sci., Paris,* **276D** : 161–164.   H. V. Merrick 1976, Recent Archaeological Research in the Plio-Pleistocene Deposits of the Lower Omo, Southwestern Ethiopia, *in* G.Ll.Isaac and E. R. McCown (eds) 1976, *Human Origins, Louis Leakey and the East African Evidence,* Menlo Park : 461–481.
8. Rich vertebrate fauna including mammals, terrestrial and aquatic reptiles and fish.   Mammalia include Proboscidea (5 spp.), Rhinocerotidae (3 spp.), Equidae (3–4 spp.), Chalicotheridae (1 sp.), Hippopotamidae (4 spp.), Suidae (at least 12 spp.), Giraffidae (4 spp.), one Camelid, Bovidae (over 20 spp.), Hyracoidea (2 spp.), one Lagomorph, Rodentia (7 families), Carnivora, Chiropterans, Galagines, and Cercopithecoidea (4 Colobinae, 3 Papionines, and 2 Theropithecines).   Y. Coppens and F. C. Howell 1974, Les faunes de mammifères fossiles de formations plio-pléistocènes de l'Omo en Ethiopie (Proboscidea, Perissodactyla, Artiodactyla), *C.r. hebd. Séanc. Acad. Sci., Paris,* **278D** : 2275–2278.   F.C. Howell and Y. Coppes 1974, Les faunes de mammifères fossiles des formations plio-pléistocènes de l'Omo en Ethiopie (Tubulidentata, Hyacoidea, Lagomorpha, Rodentia, Chiroptera, Insectivora, Carnivora, Primates), *C.r. hebd. Séanc. Acad. Sci., Paris,* **278D** : 2421–2424.   Y. Coppens 1975, Evolution des Mammifères de leur fréquences et de leurs associations, au cours du Plio-Pleistocene dans la basse vallée de l'Omo en Ethiopie, *C.r. hebd. Séanc. Acad. Sci., Paris,* **281D** : 1571–1574.
9. ────────
10. A3: 3·0–1·4 Myr.   F. C. Howell 1972, Pliocene–Pleistocene Hominidae in Eastern Africa.   Absolute and relative ages, *in* W. W. Bishop and J. A. Miller (eds) 1972, *Calibration of Hominoid Evolution,* Edinburgh : 331–368.   F. H. Brown 1972,

TABLE 2

Stratigraphic relationship of localities yielding hominids in the Shungura Formation, Omo.
Members take their designation from the underlying Tuff.
(After Howell & Coppens 1974)

Member	Tuff	Age Range Myr	Mean Myr	Hominid Locality
L				
	L	1·27–1·41	1·34	
K				Omo K7
J				F18, F203
	J			
H	I2	1·81–1·87	1·84	Omo 74
	H			
G				Omo 75, Loc 627 Loc 626 Omo 50 Omo SH1, Omo 47 Loc 105 Locs 7, 7A, 74A, Omo 29, 751, 755 Locs 726, 797 Locs 427, 754 Locs 686, 628, Omo 141 Omo 76, 136
	G	1·93	1·93	
F				Loc 238 Locs 420, 209 Loc28 Loc 157, Loc 465 F 22 Omo 76R, Omo 76 Loc 398, Omo 33, 56–7
	F	1·99–2·06	2·04	
E				Loc 40, Omo 57–5 Omo 57–4, F 933, Omo 57–6 Locs 338x, 338y Locs 26, 10 Omo 44
	E	2·12	2·12	
D				Locs 9, 64, 571 Locs 161, 704
	D	2·16–2·6	2·41	Locs 50, 296
C				Locs 51, 144, 561, 345–A, Omo 18, 84 Locs 54, 136 Locs 55, 55–S Locs 45, 849, 362, 327 Locs 183, 62 Loc 795
	C			
B				Loc 1–N
	B10	2·93–2·94	2·94	Loc 2, Omo 28 Omo 28–S
	B	3·79–4·99	4·39	

Radiometric dating of sedimentary formations in the lower Omo valley, southern Ethiopia, *in* W. W. Bishop and J. A. Miller (eds) 1972 : 273–287.   R. T. Shuey, F. H. Brown and M. K. Croes 1974, Magnetostratigraphy of the Shungura formation Southwestern Ethiopia: fine structure of the lower Matuyama polarity epoch, *Earth planet. Sci. Lett.,* **23** : 249–260.

11.   Fossil Remains listed under member and locality, with registration numbers of bones in brackets:

Member B (Upper). A2: 2·94 Myr.

2.	2 lower M.
1-N.	lower M crown.
Omo 28.*	lower M germ.
Omo 28-S.	lower M (f).

Member C.   A3: 2·5 Myr.

Omo 18.	corpus mandibulae, roots of dentes (18–18). **Holotype** of *Paraustralopithecus aethiopicus* Arambourg & Coppens, 1968. Plate 3. Also upper 2I1, 2I2, P3, 2M1; lower I1, P3, M1/2; phalanx manus.

Member E.   A3: 2·12–2·0 Myr.

10.	lower M3 (f).
26.	upper M1; lower M2.
40.	rt ulna (i) (40–19).
Omo 44.	upper P3; corpus mandibulae (f) with roots of dentes (2466).
Omo 57–4.	upper P3 (f), M3 (f); rt corpus mandibulae with P4 crown (f), M1 germ (57–41).
Omo 57–5.	P (f), 4M (f).
Omo 57–6.	upper M2 germ (f).
338X.	upper lt dm2 (f), P3 (f), P4, M2; lower C (f), P4 (f), M (f), M3 crown.
338Y.	juvenile: 2 parietalia, occipitale (i) (338Y–6); humerus diaphysis.
P.933.	upper M1.

Member F.   A3: 2·0–1·9 Myr.

(Kalam) F. 22.	lower M2 crown, M3 crown.
28.	upper P crown (f), M2; lower C, P3 germ (f), M2, M3.
Omo 33.	upper P3, M1?, M (f), M1/2, 2M3; lower P4, M2, M3, M3 germ.
Omo 76.	lower 2M3 (f), P or M (f).
Omo 76r.	lower M2 (f), M1/2 (f).
157.	lower M3.
209.	upper 3M (f); lower M1/2 (f).
238.	upper M2.
398.	upper M3; lower I (f), P3 crown, P4 germ, 2M (f), 2M1 (f), M3, M3 (f).
420.	lower P4 (f).
465.	lower P4 crown (f), P4 (f), P crown (f).

Member G.   A3: 1·9–1·84 Myr.

Omo SH-1.	upper M1 crown.
7.	lower M2.

*The prefix *Omo* indicates the discovery was made by the French expedition.   The number/letter designation following refers to the locality.   Localities situated near the Kalam police post are so indicated.

7A.	mandibula with all dentes except 31 (7A-125).
Omo 29.	lower P3.
Omo 47.	upper M (f); lower M1, M2, M3 (f).
Omo 50.	upper I2.
74-A.	rt corpus mandibulae with C and P4 (74-A-21).
Omo 75.	associated dentes: upper 2P4 (f), 2M1 crown (f), 2M2 crown (f), M3 (i); rt corpus mandibulae with 2P3, P4, 2M1 (f), 2M2, M3 (75-14a, b).
Omo 75i.	lower P3.
Omo 75s.	lower I2 or C, M1, M3 germ; proximal radius (75-s-1317).
105.	vertebra lumbalis (105–7).
Omo 136.	upper M (f); lower M3 germ, M3 crown.
Omo 141.	upper M1/2 (f), M3 (f).
427.	rt corpus mandibule (f) with erupting P3, P4 in crypt, roots of dm2 and M1, M2 (i) (427–7).
626.	upper I1 crown.
627.	lt frontale (f)—supraorbital region (627–236).
628.	upper I1, P3 (f), 2P4, M (f); lower P4, M1, M2, M3, M3 (f), M crown (f).
686.	lower I2.
726.	upper P3 crown (f).
754.	femur diaphysis (f) (754–8).
797.	lower P4 (f).

Members H, J and
later occurrences.   A3: 1·84–1·5 Myr.

(Kalam) F. 203.	lower M3 crown.
Omo K-7.	lower M2 crown.
(Kalam) F. 18.	upper I2.
Omo 74.	lower P4 germ.

12. C. Arambourg and Y. Coppens 1967, Sur la découverte dans le Pléistocène inférieur de la vallée de l'Omo (Ethiope) d'une mandibule d'australopithécien, *C.r. hebd. Séanc. Acad. Sci. Paris,* **265 D** : 589–590.   F. C. Howell 1968, Omo Research Expedition, *Nature, Lond.,* **219** : 567–572.   F. C. Howell 1969, Remains of Hominidae from Pliocene/Pleistocene formations in the lower Omo basin, Ethiopia, *Nature, Lond.,* **223** : 1234–1239. Y. Coppens 1970, Les restes d'Hominidés des séries inférieurs et moyennes des formations Plio-Villafranchiennes de l'Omo en Ethiopie, *C.r. hebd. Séanc. Acad. Sci., Paris,* **271 D** : 2286–2289. Y. Coppens 1971, Les restes d'Hominidés des séries supérieurs des formations Plio-Villafranchiennes de l'Omo en Ethiopie, *C.r. hebd. Séanc. Acad. Sci., Paris,* **272 D** : 36–39.   Y. Coppens 1973a, Les restes d'Hominidés des séries inférieures et moyennes des formations Plio-Villafranchiennes de l'Omo en Ethiopie (récoltes 1970, 1971 et 1972), *C.r. hebd. Séanc. Acad. Sci., Paris,* **276 D** : 1823–1826.   Y. Coppens 1973b, Les restes d'Hominidés des séries inférieures et moyennes/supérieures des formations plio-villafranchiennes de l'Omo en Ethiopie (récoltes 1970, 1971 et 1972), *C.r. hebd. Séanc. Acad. Sci., Paris,* **276 D** : 1981–1984.

13. C. Arambourg and Y. Coppens 1968, Découverte d'un australopithécien nouveau dans les gisements de l'Omo (Ethiopie), *S. Afr. J. Sci.,* **64** : 58–59.   Holotype of *Paraustralopithecus aethiopicus*, Arambourg & Coppens.   F. C. Howell and Y. Coppens

      1973, Deciduous teeth of Hominidae from the Pliocene/Pleistocene of the lower Omo basin, Ethiopia, *J. hum. Evol.*, **2** : 461–472.

14. ————

15. F. C. Howell 1969.   Y. Coppens 1970, 1971, 1973a, 1973b.   F. C. Howell and Y. Coppens 1973.

16. F. C. Howell 1971, Hominidae: *Australopithecus, McGraw-Hill Yearbook of Science and Technology,* 1970 : 220–222.   Y. Coppens 1971, Les australopithèques rehabilités.   *Sciences et Avenir* (N° special: La Vie Préhistorique): 80–83.   F. C. Howell 1972, Recent advances in human evolutionary studies, *in* S. L. Washburn and P. C. Dolhinow (eds) 1972, *Perspectives in Human Evolution,* vol. 2 : 51–128.   F. C. Howell and Y. Coppens 1974, Inventory of remains of Hominidae from Pliocene/Pleistocene formations of the lower Omo basin, Ethiopia (1967–1972), *Am. J. phys. Anthrop.*, **40** : 1–16.   Y. Coppens 1975, Evolution des Hominidés et de leur environnement au cours du Plio-Pléistocène dans la basse vallée de l'Omo en Ethiopie, *C.r. hebd. Seanc. Acad. Sci., Paris,* **281** : 1693–1696.

17. National Museum, Addis Ababa.

18. Department of Anthropology, University of California, Berkeley, California, 94720, U.S.A.

1. Usno.

2. Exposures in NE quadrant of lower Omo basin, 15 km S of junction of Usno stream with Omo River.   5° 18–20′ N, 36° 12′ E.

3. Expedition under the direction of F. C. Howell 1967, 1968.

4. Fossiliferous sands interbedded with silts: Brown and White sands of Usno formation.   J. de Heinzelin and F. H. Brown 1969, some early Pleistocene deposits of the lower Omo valley: the Usno formation, *Quaternaria,* **11** : 31–46.

5. ————

6. Middle Pliocene: equivalent to Shungura member B.   F. C. Howell 1972, Pliocene/Pleistocene Hominidae in Eastern Africa: absolute and relative ages, *in* W. W. Bishop and J. A. Miller (eds) 1972, *Calibration of Hominoid Evolution,* Edinburgh: 331–368.

7. None reported.

8. Rich vertebrate fauna including mammals, terrestrial and aquatic reptiles, and fish.   Forty species of Mammalia include *Nyanzachoerus, Notochoerus capensis, Sivatherium, Tragelaphus nakuae, Elephas recki.*   F. C. Howell, L. S. Fichter and G. Eck 1969, Vertebrate assemblages from the Usno formation, White Sands and Brown Sands localities, lower Omo basin, Ethiopia, *Quaternaria,* **11** : 65–88.   F. C. Howell and Y. Coppens 1974, Les faunes de mammifères fossiles de formations plio/pléistocènes de l'Omo en Ethiopie, (Proboscidea, Perissodactyla, Artiodactyla), *C.r. hebd. Séanc. Acad. Sci. Paris,* **278D** : 2275–2278.   F. C. Howell and Y. Coppens 1974, Les faunes de mammifères fossiles des formations Plio-Pléistocène de l'Omo en Ethiopie (Tubulidentata, Hyacoidea, Lagomorpha, Rodentia, Chiroptera, Insectivora, Carnivora, Primates), *C.r. hebd. Séanc. Acad. Sci., Paris,* **278D** : 2421–2424.

9. ————

10. A3: 2·9 Myr on basis of K/Ar date of 2·64–2·97 Myr of underlying triple tuff.   F. H. Brown 1972, Radiometric dating of sedimentary formations in the lower Omo valley,

southern Ethiopia, *in* W. W. Bishop and J. A. Miller (eds) 1972, *Calibration of Hominoid Evolution*, Edinburgh : 273–287.

11. **B4.**   upper lt P4.
     B14a.   upper lt P4.
     B14b.   upper lt P4.
     B20.   upper dm2 (f).
     B23a.   upper lt P4.
     B23b.   upper lt M1.
     B27.   upper rt I1.
     B28.   lower dm2 (f).
     B39a.   upper lt P3.
     B39b.   upper lt P4.
     B39c.   upper M1 (f).
     W23.   lower lt P4.
     W508.   lower rt M1.
     W749.   upper rt M2.
     W750.   upper P3 (ff).
     W751.   lower P4 (f).
     W752.   lower rt M1.
     W753.   upper rt dml.
     W758.   lower lt M1 (f).
12. F. C. Howell 1968, Omo Research Expedition, *Nature, Lond.,* **219** : 567–572.
13. F. C. Howell 1969, Hominid teeth from White Sands and Brown Sands localities, lower Omo basin (Ethiopia), *Quaternaria,* **11** : 47–64. Cf. *Australopithecus africanus* (but 2 dm are more robust).   F. C. Howell and Y. Coppens 1973, Deciduous Teeth of Hominidae from the Pliocene/Pleistocene of the Lower Omo Basin, Ethiopia, *J. hum. Evol.,* **2** : 461–472.
14. ————
15. F. C. Howell 1969.   F. C. Howell and Y. Coppens 1973.
16. ————
17. National Museum, Addis Ababa.
18. Department of Anthropology, University of California, Berkeley, California, U.S.A.

KENYA

Diana Margaret LEAKEY

South African Museum,
P.O. Box 61,
Cape Town,
South Africa.

&

The Editors.

BARINGO

1. Baringo (Kapthurin).
2. *In situ* and on erosion slope about 1·5 km W of Marigat–Nginyang road where it crosses the Kapthurin River, W of Lake Baringo.   0° 47′ N, 36° 05′ E.
3. E. Kandini, assistant to M. D. Leakey, 7 February 1966 (KNM–BK67); assistants to R. E. F. Leakey and D. M. Leakey, 1966 (KNM–BK63–66).
4. Kapthurin Beds, sandy clays *in situ* (KNM-BK63–65); surface about 13·5 m (45 ft) below base of tuff which caps the plateau (KNM-BK66 & 67).   J. E. Martyn 1969, *in* M. Leakey 1969, An Acheulean Industry with Prepared Core Technique and the Discovery of a Contemporary Hominid Mandible at Lake Baringo, Kenya, *Proc. prehist. Soc.,* **35** : 48–76.
5. ————————
6. Middle Pleistocene.
7. Acheulian with prepared cores.   M. Leakey 1969.
8. *Colobus, Thryonomys, Potamochoerus, Phacochoerus, Hippopotamus amphibius, Tragelaphus spekei, Cephalophus* cf. *harveyi.*   R. E. F. Leakey 1969, *in* M. Leakey 1969.
9. Hominid ulna: eU308 = 10 ppm.
10. ————————
11. **BK63.**   rt metatarsale I.
    BK64.   phalanx.
    BK65.   phalanx.
    BK66.   rt ulna (f).
    BK67.   young adult: mandibula, lt M2-3, roots P3-M1, alveoli I1-C, rt alveoli I1–C, roots P3, P4 (f), M1 (f), M2–3.
12. M. Leakey 1969.
13. P. V. Tobias 1969, *in* M. Leakey 1969 (KNM-BK67).   *Homo* cf. *erectus.*
14. ————————
15. ————————
16. ————————
17. National Museum, P.O. Box 40658, Nairobi.
18. National Museum, P.O. Box 40658, Nairobi.

1. Chemeron.
2. On erosion slope at site JM85 in the valley of the Kapthurin River, W of Lake Baringo.   0° 35′ N, 35° 55′ E.
3. J. Kimengich, assistant to J. E. Martyn, October 1965, whilst working for the East African Geological Research Unit, Bedford College, London.
4. Surface, probably from a very thin grit band about 2·5 m (8 ft) from the base of the Upper Fish Beds in the Chemeron Formation.   J. E. Martyn 1967, Pleistocene Deposits and New Fossil Localities in Kenya, *Nature, Lond.,* **215** : 476–480.
5. ————————
6. Lower Pleistocene.
7. ————————

8.   Abundant remains of Pisces: *Clarias, Tilapia.*
9.   ———
10. ———
11. **BC 1.**   rt temporale (f).
12. J. E. Martyn 1967.
13. P. V. Tobias *in* J. E. Martyn 1967.   Hominidae indet.
14. ———
15. P. V. Tobias *in* J. E. Martyn 1967.
16. ———
17. National Museum, P.O. Box 40658, Nairobi.
18. National Museum, P.O. Box 40658, Nairobi.

1.   Chesowanja.
2.   On erosion slope 1·5 km W of the track from Mukutau to Chemoigut, E of Lake
     Baringo.   0° 39′ N, 36° 12′ E.
3.   L. Rungu, assistant to J. Carney, 1970, whilst working for the East African Geological
     Research Unit, Bedford College, London (CH1); M. Pickford and A. Hill, August 1973
     (CH 302).
4.   Surface, *ex* Yellow Marls of the Chemoigut Beds.   J. Carney, A. Hill, J. A. Miller and
     A. Walker 1971, Late Australopithecine from Baringo District, Kenya, *Nature, Lond.,*
     230 : 509–514.
5.   ———
6.   Lower Pleistocene.
7.   Pebble tools and flakes.   W. W. Bishop, M. Pickford and A. Hill 1975, New evidence
     regarding the Quaternary geology, archaeology and hominids of Chesowanja, Kenya,
     Nature, Lond., **258** : 204–208.
8.   *Equus, Ceratotherium, Pronotochoerus jacksoni, Hippopotamus* cf. *amphibius, Elephas*
     cf. *recki, Deinotherium bozasi.*   W. W. Bishop, G. R. Chapman, A. Hill and J. A.
     Miller 1971, Succession of Cainozoic Vertebrate Assemblages from the Northern
     Kenya Rift Valley, *Nature, Lond.,* **233** : 389–394.
9.   ———
10. A3: 1·1 or 1·2 Myr on the basis of K/Ar dating of trachyte from the Chepchuk
     volcanics.   J. Carney *et al.* 1971.
11. **CH1.**   cranium (ff), rt C–M3.
     CH302.   M (ff).
12. J. Carney *et al.* 1971 (CH1).
13. A. Walker 1971, *in* J. Carney *et al.* 1971.   *Australopithecus* (CH1).
14. ———
15. J. Carney *et al.* 1971 (CH1).
16. ———
17. National Museum, P.O. Box 40658, Nairobi.
18. National Museum, P.O. Box 40658, Nairobi.

BROMHEAD'S SITE

See under **ELMENTEITA**

CHEMERON

See under **BARINGO.**

CHESOWANJA

See under **BARINGO.**

EAST RUDOLF

See under **ILERET and KOOBI FORA**

ELMENTEITA

1. Elmenteita (Bromhead's Site).
2. Buried in a cliff at Bromhead's Site on the S bank of the Makalia River, S of Lake Nakuru in Elmenteita District on a farm formerly owned by Mr Monroe.   0° 29′ S, 36° 05′ E.
3. W. S. Bromhead, 1917; L. S. B. Leakey and B. S. Newsam, 1926 and 1927, during the first East African Archaeological Research Expedition.
4. In stratified silts banked against a cliff of volcanic ash.   J. D. Solomon 1971, *in* L. S. B. Leakey 1931, *Stone Age Cultures of Kenya Colony*, Cambridge : 254.
5. Burials mostly washed to the bottom of the gorge by rising lake water but with some protected *in situ* by large boulders.   Traces of one almost complete burial were found.   L. S. B. Leakey 1935, *Stone Age Races in Kenya*, London : 59.
6. Holocene.
7. Elmenteitan.   L. S. B. Leakey 1931.
8. *Papio anubis neumanni, Procavia, Phacochoerus aethiopicus africanus, Bos.*   A. T. Hopwood 1931, *in* L. S. B. Leakey 1931 : 271.
9. Skull A: F = 1·3%, 100F/P205 = 5·4, eU308 = 16 ppm, N = 3·66%.
10. A1 : 7410 ± 160 BP (UCLA-1757) on the basis of C14 dating of human bone.
11. **Elmenteita A–Q.**   adults and juveniles: crania and calottes, some with associated mandibulae.
11. **Elmenteita 1–24.**   adults and juveniles: isolated mandibulae, post-cranial bones.
12. Anon. 5 May 1927, *The Times*.
13. L. S. B. Leakey 1935.   *Homo sapiens*.
14. G. P. Rightmire 1975, New Studies of Post-Pleistocene Human Skeletal Remains from the Rift Valley, Kenya, *Am. J. phys. Anthrop.,* **42** : 351–370.
15. L. S. B. Leakey 1935.

16. L. S. B. Leakey 1927, Stone Age Man in Kenya Colony, *Nature, Lond.,* **120** : 85–86.
17. British Museum (Natural History), Cromwell Road, London SW7 5BD, England.   Reg. Nos EM 808–1007.
18. University Museum, University of Pennsylvania, 33rd & Spruce Streets, Philadelphia 4, Pennsylvania, U.S.A.

FORT TERNAN*

1. Fort Ternan.
2. In excavation on land formerly owned by F. Wicker at Fort Ternan, 64 km E of Kisumu on Lake Victoria.   0° 13′ S, 35° 20′ E.
3. H. Mukiri and other assistants to L. S. B. Leakey, 1961 (KNM-FT46), 1962 (KNM-FT45), 1967 (KNM-FT7), not known (KNM-FT3318).
4. Tuffs interbedded with lava flows, Bed 6 or 7.   W. W. Bishop and F. Whyte 1962, Tertiary Mammalian Faunas and Sediments in Karamoja and Kavirondo, East Africa, *Nature, Lond.,* **196** : 1283–1287.
5. ————
6. Middle Miocene (equivalent to Vindobonian).   W. W. Bishop and J. A. Miller (eds) 1972, *Calibration of Hominoid Evolution,* Edinburgh.
7. ————
8. *Leakeymys ternani, Dryopithecus* cf. *africanus, Dryopithecus* cf. *nyanzae,* cf. *Limnopithecus legetet, Gomphotherium, Paradiceros mukirii, Palaeotragus primaevus, Oioceros tanyceras, Protragocerus labidotus.*   R. Lavocat 1964, Fossil Rodents from Fort Ternan, Kenya, *Nature, Lond.,* **202** : 1131.   C. S. Churcher 1970, Two New Upper Miocene Giraffids from Fort Ternan, Kenya, East Africa *in* L. S. B. Leakey and R. J. G. Savage (eds) 1970, *Fossil Vertebrates of Africa,* Vol. 2: 1–105.   A. W. Gentry 1970, The Bovidae (Mammalia) of the Fort Ternan Fossil Fauna, *in* L. S. B. Leakey and R. J. G. Savage (eds) 1970 : 243–323.   P. J. Andrews and A. C. Walker 1976, The Primate and Other Fauna from Fort Ternan, Kenya, *in* G. Ll. Isaac and E. R. McCown (eds) 1976, *Human Origins,* Menlo Park: 279–304.   D. A. Hooijer 1968, A Rhinoceros from the Late Miocene of Fort Ternan, Kenya, *Zoöl. Meded. Leiden,* **43** : 77–92.
9. ————
10. A3: *c.* 14–12·5 Myr on basis of K/Ar dating of underlying tuff and overlying phonolite.   J. F. Evernden, D. E. Savage, G. H. Curtis and G. T. James 1962, Potassium-argon dates and the Cenozoic mammalian chronology of North America, *Am. J. Sci.,* **262** : 145–198.   14·7 ± 0·7 and 14·0 ± 0·2 Myr on basis of K/Ar dating of biotite from Bed 5.   12·5 ± 0·4 and 12·6 ± 0·7 Myr on basis of ^{40}Ar/^{39}Ar dating of phonolite 22 m (70 ft) above Bed 6–7.   W. W. Bishop, J. A. Miller and F. J. Fitch 1969, New potassium-argon age determinations relevant to the Miocene Fossil mammal sequence in East Africa, *Am. J. Sci.,* **267** : 669–699.   P. J. Andrews and A. C. Walker 1976.
11. **FT7.**   rt corpus mandibulae (f), P4, M1.

*Contributed by P. J. Andrews.

FT8.   lt upper maxilla (f), C (unerupted).
FT45.   lt corpus mandibulae (f), P3, P4, mesial roots of M1, alveoli of I1–C.
FT46/47. lt maxilla (f), C, P4–M2, roots of P3; rt maxilla (f), M1, M2, roots of
             M3.   **Holotype** of *Kenyapithecus wickeri* Leakey, 1962.   Plate 4.
FT48.   lower rt M2.
FT3318.   lower rt C.
FT45, 46/47 are thought to be one individual.

12.   L. S. B. Leakey 1962, A new Pliocene fossil primate from Kenya, *Ann. Mag. nat. Hist.*
      (13), **4** : 689–696 (FT46 & 47).   L. S. B. Leakey 1968, Upper Miocene Primates from
      Kenya, *Nature, Lond.,* **218** : 527–528 (FT45).
13.   L. S. B. Leakey 1962 (FT46 & 47).   *Kenyapithecus wickeri.*   P. J. Andrews 1971, *Ra-*
      mapithecus wickeri Mandible from Fort Ternan, Kenya, *Nature, Lond.,* **231** : 192–194
      (FT45).   *Ramapithecus wickeri.*
14.   A. C. Walker and P. J. Andrews 1973, Reconstruction of the dental arcades of *Ra-*
      mapithecus wickeri, Nature, Lond., **244** : 313–314.
15.   B. G. Campbell 1972, Conceptual Progress in Physical Anthropology: Fossil Man,
      Ann. Rev. Anthrop., **1** : 27–54 (FT45, 46 & 47).   A. C. Walker and P. J. Andrews
      1976 (FT45, 47).
16.   E. L. Simons 1964, On the Mandible of *Ramapithecus, Proc. natn. Acad. Sci. U.S.A.,*
      51 : 528–531.   E. L. Simons and D. R. Pilbeam 1965, Preliminary Revision of the
      Dryopithecinae (Pongidae, Anthropoidea), *Folia Primatol.,* **3** : 81–152.   D. R.
      Pilbeam and E. L. Simons 1965, Some Problems of Hominid Classification, *Am.*
      Scient., **53** : 237–259.   E. L. Simons 1968, Late Miocene Hominid from Fort Ternan,
      Kenya, *Nature, Lond.,* **221** : 448–451.   L. S. B. Leakey 1969, Fort Ternan Hominid,
      Nature, Lond., **222** : 1202.
17.   National Museum, P.O. Box 40658, Nairobi.
18.   National Museum, P.O. Box 40658, Nairobi.   Wenner-Gren Foundation for
      Anthropological Research, 14 East 71st Street, New York, N.Y. 10021, U.S.A.   (FT46
      & 47).

GAMBLE'S CAVE

1.   Gamble's Cave II.
2.   In rock-shelter in a cliff facing a farm called Miti Mingi, SW of Lake Nakuru in
     Elmenteita District.   0° 33′ S, 36° 05′ E.
3.   L. S. B. Leakey and other members of the first and second East African Archaeological
     Expeditions, 28–29 June 1927, 4 July 1927, 18 July 1928 and 14 March 1929.
4.   Layer 11, between the 2nd and 3rd occupation levels, 4·5–5·0 m below the modern sur-
     face.   L. S. B. Leakey 1931, *Stone Age Cultures of Kenya Colony,* Cambridge : 116.
5.   Bodies ultra-contracted, red-ochred and placed under stones on the 3rd occupation level
     (layer 12).   L. S. B. Leakey 1935, *Stone Age Races of Kenya,* London : 48.
6.   Holocene.
7.   Late Upper Kenya Capsian C.   L. S. B. Leakey 1931 : 109.
8.   Recent fauna.
9.   Gamble's Cave 4 : F = 2·4%, 100F/P205 = 8·4, eU308 = 7 ppm, N = 0·1%.
     Second sample: N = 1·44% (UCLA–1756).

10. A1: 8210 $\pm$ 260 BP (UCLA-1756) on the basis of C14 dating of bone from Gamble's
     Cave 4.
11. **Gamble's Cave 1.**   maxilla, malar, mandibula, 2 metatarsalia (f), patella (f), tibia (ff), 2
                            femora (ff).
     Gamble's Cave 2.   ramus mandibulae, dentes.
     Gamble's Cave 3.   occipitale (ff), M, tibia (ff), 2 femora (ff), calcaneus (ff).
     Gamble's Cave 4.   adult, 20–25 years male ?: cranium, mandibula, skeleton in matrix
                            (f).
     Gamble's Cave 5.   immature male: cranium, mandibula, skeleton (f).
12. L. S. B. Leakey 1931.
13. L. S. B. Leakey 1935.   *Homo sapiens.*
14. G. P. Rightmire 1975, New Studies of Post-Pleistocene Human Skeletal Remains from
     the Rift Valley, Kenya, *Am. J. phys. Anthrop.,* **42** : 351–370.
15. L. S. B. Leakey 1935.
16. ─────────
17. British Museum (Natural History), Cromwell Road, London SW7 5BD,
     England.   Reg. Nos. EM 764–790.
18. University Museum, University of Pennsylvania, 33rd & Spruce Streets, Philadelphia 4,
     Pennsylvania, U.S.A.

ILERET and KOOBI FORA–General Entry

The following entry was prepared by Margaret Leakey from the files of the Kenya National
Museum and includes material found up to the end of 1975.   An illustrated catalogue of the
fossils is in press: R. E. F. Leakey and M. G. Leakey (eds) 1977, *Koobi Fora: Researches into
Geology, Paleontology and Human Origins.*   Vol. 1: *The hominid fossils and an introduction
to their context,* Oxford.

1. Ileret and Koobi Fora.
2. Open sites covering about 15–20 × 40 square miles between Ileret and Aliya Bay on the
    East side of Lake Turkana.   Within each region areas of outcrops have been given
    numerical designations.   (See Fig. 1).   They fall within the area 3° 35' N – 4° 19' N,
    and 36° 14' E – 36° 25' E.   In the individual entries the latitude and longitude refer to
    the central point of the area concerned.
3. Koobi Fora Research Project led by G. Ll. Isaac and R. E. F. Leakey,
    1967–1975.   (This edition of the Catalogue reports only specimens found up to the end
    of 1975.)
4. All specimens derive from the Koobi Fora formation, except ER999 which derives from
    the later Guomde Formation.   The correlations of the different members of the Koobi
    Fora formations are shown in Table 3.   Work is proceeding and these may be subject
    to further revision.
    A. K. Behrensmeyer 1970, Preliminary geological interpretation of a new hominid site
    in the Lake Rudolf basin, *Nature, Lond.,* **226** : 225–226.   C. F. Vondra, G. D.
    Johnson, B. E. Bowen and A. K. Behrensmeyer 1971, Preliminary stratigraphical
    studies of the East Rudolf basin, *Nature, Lond.,* **231** : 245–248.   R. E. F. Leakey

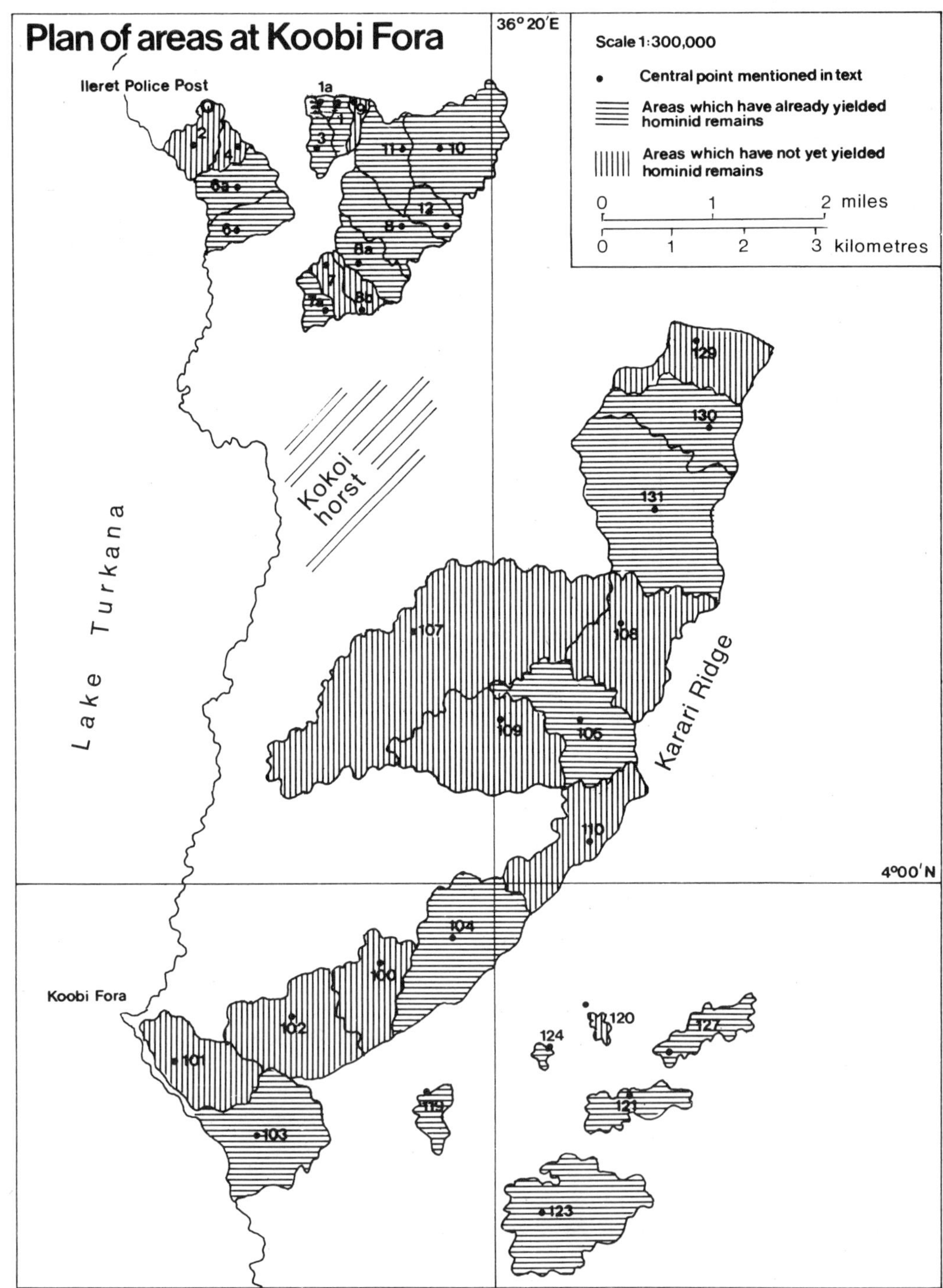

Fig. 1.   Plan of areas surveyed east of Lake Turkana.

TABLE 3

Excavated sites in the Koobi Fora region in relation to stratigraphic markers and selected hominid fossil finds. Karari Industry sites are shown in bold face and hominid fossil finds in italic. KBS Industry sites are shown thus: FxJj 1 and so on.   (From *Nature, Lond., 262.*)

Koobi Fora Coast Areas 102, 103	KBS Ridge Areas 105, 118	Okote Tuff Complex including BBS Tuff	Karari Escarpment Areas 129, 130, 131	Ileret Areas 1–12	Lower/Middle Tuff Complex	Karari Industry / KBS Industry	Loxodonta africana zone / Metridiochoerus andrewsi zone / Mesochoerus limnetes zone	Upper Member / Lower Member
			Karari Tuff	Chari Tuff				
Erosion								
For example KNM-ER *1807*				For example KNM-ER *731, 739, 992, 741, 805, 993*			*Loxodonta africana* zone	Upper Member
			KNM-ER *3230* **FxJj 20 M, E, AB** **FxJj 19** **FxJj 16, 18** **FxJj 11**	For example KNM-ER **FwJj 1** *807* **FwJj 1** *803, 818 808, 1463 729*	Lower/Middle Tuff Complex	Karari Industry		
Koobi Fora Tuff Complex	Erosion							
For example KNM-ER *730, 736, 737, 1808*			**FxJj 38** KNM-ER *1805, 1806*	For example KNM-ER *406, 407, 732, 815, 1592*			*Metridiochoerus andrewsi* zone	
	KNM-ER *405, 738, 1476, 1477, 1478, 1479*		FxJj 15					
KBS Tuff Complex	FxJj 1, 3, 10		FxJj 14	KBS Tuff Complex		KBS Industry	*Mesochoerus limnetes* zone	Lower Member
	FxJj 13		For example KNM-ER *1470, 1472, 1473, 1481, 1482, 1483*	For example KNM-ER *1590, 1593*				

1972a, Further evidence of Lower Pleistocene hominids from East Rudolf, North Kenya 1971, *Nature Lond.*, **237** : 264–269.   V. J. Maglio 1972, Vertebrate faunas and chronology of hominid-bearing sediments east of Lake Rudolf, Kenya, *Nature, Lond.*, **239** : 379–385.   B. E. Bowen and C. F. Vondra 1973, Stratigraphic relationships of the Plio-Pleistocene depositions, East Rudolf, Kenya, *Nature, Lond.*, **242** : 391–393.   Y. Coppens, F. C. Howell, G. Ll. Isaac and R. E. F. Leakey (eds) 1976, *Earliest Man and Environments in the Lake Rudolf Basin; Stratigraphy, Paleoecology and Evoluton*, Chicago: 1–615.

5. ———

6. Upper Pliocene to Lower Pleistocene.   3 faunal zones are distinguished.   See Table 3 and item 8 under individual entries.

7. Two series of archaeological occurrences can be recognized.   The older series occurs at the top of the Lower Member of the Koobi Fora Formation, mainly in the KBS Tuff complex.   The artifacts from these sites can be tentatively termed the KBS Industry. The younger contrasting series has been discovered in the Upper Member and is designated the Karari Industry (see Table 3).   So far only ER 1805, 1806 and 3230 have been found in close conjunction with artifacts.   J. W. K. Harris and G. Isaac 1976, The Karari Industry: Early Pleistocene archaeological evidence from the terrain east of Lake Turkana, Kenya, *Nature, Lond.,* **262**: 102–107.   M. D. Leakey 1970, Early artifacts from the Koobi Fora area, *Nature, Lond.,* **226** : 228–230.   G. Ll. Isaac, R. E. F. Leakey and A. K. Behrensmeyer 1971, Archaeological traces of early hominid activities, east of Lake Rudolf, Kenya, *Science, N.Y.,* **173** : 1129–1134.

8. Four faunal zones have been established, three of which are relevant to the hominid-bearing deposits.   See Table 3.   V. J. Maglio 1971, Vertebrate Faunas from the Kubi Algi, Koobi Fora and Ileret Areas, East Rudolf, Kenya, *Nature, Lond.,* **231** : 248–249.   V. J. Maglio 1972.   Y. Coppens *et al.* 1976.

9. ———————

10. A3: 3·18 ± 0·09 Myr on basis of 40Ar/39Ar dating of sanidine crystals from Tulu Bor Tuff.   F. J. Fitch and J. Miller 1976, *in* Y. Coppens *et al.* (eds) 1976: 123–47.
A3: 2·61 ± 0·26 Myr (Leakey IB1 and IB2) on basis of K/Ar and 40Ar/39Ar dating of the KBS Tuff which marks the upper limit of the Lower Member.   F. J. Fitch and J. A. Miller 1970, Radioisotopic age determinations of Lake Rudolph artifact site, *Nature, Lond.,* **226**: 226–228.   F. J. Fitch, I. C. Findlater and R. T. Watkins 1974, Dating of the rock succession containing fossil hominids on East Rudolf, Kenya, *Nature, Lond.,* **251** : 213–215.   Conventional K/Ar dates on pumice from KBS Tuff distinguish 2 units.   Area 10 and Area 105: 1·6 ± 0·05 Myr; Area 121: 1·82 ± 0·04 Myr.   G. H. Curtis, Drake T. Cerling and J. H. Hampel 1975, Age of KBS Tuff in Koobi Fora Formation, East Rudolf, Kenya, *Nature, Lond.,* **258**: 395–398.   2·44 ± 0·08 Myr on basis of fission-track dating of pumice from the KBS Tuff.   A. J. Hurford, A. J. W. Gleadow and C. W. Naeser 1975, Fission-track dating of pumice from the KBS Tuff, East Rudolph, Kenya, *Nature, Lond.,* **263**: 738–740.   F. J. Fitch, P. J. Hooker and J. A. Miller 1976, 40Ar/39Ar dating of the KBS Tuff in Koobi Fora Formation, East Rudolph, Kenya, *Nature, Lond.,* **263**: 740–744.   Palaeomagnetic investigation revealed reversed polarity below the KBS Tuff and normal polarity above the KBS Tuff.   A. Brock and G. Ll. Isaac 1974, Palaeomagnetic stratigraphy and chronology of hominid-bearing sediments east of Lake Rudolph, Kenya, *Nature, Lond.,* **247**: 344–348.
A3: 1·32 ± 0·01 Myr on basis of 40Ar/39Ar dating of sanidine crystals from Karari Tuff.   F. J. Fitch and J. Miller 1976, *in* Y. Coppens *et al.* (eds) 1976: 123–147.

11–15.   See under individual entries.

16. H. M. McHenry 1974, How large were the Australopithecines? *Am. J. phys. Anthrop.,* **40** : 329–340.   (ER 736, 737, 738, 739, 741).   A. K. Behrensmeyer 1975, The taphonomy and paleoecology of Plio-Pleistocene vertebrate assemblages East of Lake Rudolf, Kenya, *Bull. Mus. comp. Zool. Harv.,* **146** : 473–578.   G. Ll. Isaac and E. R. McCown (eds) 1976, *Human Origins,* Menlo Park: 1–591.   C. Jolly (ed.) 1977, *Early Hominids of Africa,* London.

17. National Museum, P.O. Box 40658, Nairobi.

18.   National Museum, P.O. Box 40658, Nairobi.

The entries are grouped according to the areas from which they were recovered and the areas themselves are given numerically, those of the Ileret region coming first.   The following list of hominids in numerical order is intended to facilitate the location of individual entries. Museum Registration numbers are as listed but with the prefix KNM; thus KNM – ER164 etc.

Reg. No.	Area	Element	Page
ER164	Koobi Fora area 104	parietale, vertebrae, phalanges	51
ER403	Koobi Fora area 103	mandibula	50
ER404	Ileret area 7A	mandibula	46
ER405	Koobi Fora area 105	palate	52
ER406–7	Ileret area 10	cranium; calvaria	48
ER417	Koobi Fora area 129	parietale	56
ER725	Ileret area 1	mandibula	43
ER726	Ileret area 10	mandibula	48
ER727	Ileret area 6A	mandibula	45
ER728	Ileret area 1	mandibula	43
ER729	Ileret area 8	mandibula	47
ER730	Koobi Fora area 103	mandibula	50
ER731	Ileret area 6A	mandibula	45
ER732	Ileret area 10	cranium	48
ER733	Ileret area 8A	maxilla, mandibula	47
ER734	Koobi Fora area 103	parietale	50
ER736–7	Koobi Fora area 103	femur, femur	50
ER738	Koobi Fora area 105	femur	52
ER739	Ileret area 1	humerus	43
ER740	Ileret area 3	humerus	44
ER741	Ileret area 1	tibia	43
ER801–2	Ileret area 6A	mandibula, dentes	45
ER803	Ileret area 8A	ossa longa	47
ER805	Ileret area 1	mandibula	43
ER806–9	Ilcrct area 8A	dentes; maxilla	47
ER810–4	Koobi Fora area 104	mandibula; parietale; talus, tibia	51
ER815	Ileret area 10	femur	48
ER816	Koobi Fora area 104	dentes	51
ER817	Koobi Fora area 124	mandibula	55
ER818	Ileret area 6A	mandibula	45
ER819–20	Ileret area 1	mandibulae	43
ER992	Ileret area 3	mandibula	44
ER993	Ileret area 1	femur	43
ER997–8	Koobi Fora area 104	metatarsale; I	51
ER999	Ileret area 6A	femur	45
ER1170–1	Ileret area 6A	cranial fragments; dentes	45
ER1462	Koobi Fora area 130	M3	56
ER1463	Ileret area 1	femur	43
ER1464	Ileret area 6A		45
ER1465	Ileret area 11	femur	49
ER1466	Ileret area 6	parietale	45
ER1467	Ileret area 3	M3	44
ER1468	Ileret area 8	mandibula	47

Reg. No.	Area	Element	Page
ER1469–75	Koobi Fora area 131	mandibula; cranium; tibia; femora; humerus	56
ER1476–80	Koobi Fora area 105	talus, tibiae; mandibula; dentes	52
ER1481–3	Koobi Fora area 131	ossa longa, mandibulae	56
ER1500	Koobi Fora area 130	skeleton (ff)	56
ER1501–5	Koobi Fora area 123	mandibula, femora, humerus	54
ER1506	Koobi Fora area 121	mandibula	54
ER1507–8	Koobi Fora area 127	mandibula, dens	55
ER1509	Koobi Fora area 119	dentes	54
ER1510	Koobi Fora area 119		54
ER1515	Koobi Fora area 103	I	50
ER1590–3	Ileret area 12	cranium, dentes, humerus, femur	49
ER1648	Koobi Fora area 105	cranial fragments	52
ER1800	Koobi Fora area 130	cranium	56
ER1801	Koobi Fora area 131	mandibula	56
ER1802	Koobi Fora area 131	mandibula	56
ER1803	Koobi Fora area 131	mandibula	56
ER1804	Koobi Fora area 104	maxilla	51
ER1805	Koobi Fora area 130	cranium, mandibula	56
ER1806	Koobi Fora area 130	mandibula	56
ER1807	Koobi Fora area 103	femur	50
ER1808	Koobi Fora area 103	cranium, mandibula, dentes, femur	50
ER1809	Koobi Fora area 127	femur	55
ER1810	Koobi Fora area 123	tibia	54
ER1811	Koobi Fora area 123	tibia	54
ER1812	Koobi Fora area 123	mandibula	54
ER1813	Koobi Fora area 123	cranium	54
ER1814	Koobi Fora area 127	maxilla	55
ER1815	Ileret area 1	talus	43
ER1816	Ileret area 6A	mandibula	45
ER1817	Ileret area 1	mandibula	43
ER1818	Ileret area 6A	I	45
ER1819	Ileret area 8	M3	44
ER1820	Koobi Fora area 103	mandibula	50
ER1821	Koobi Fora area 123	parietale	54
ER1822	Koobi Fora area 123	femur	54
ER1823	Ileret area 6A	metatarsale	45
ER1824	Ileret area 6A	humerus	45
ER1825	Ileret area 6A	atlas	45
ER2592	Ileret area 6	parietale	45
ER2593	Ileret area 6	M	45
ER2594	Ileret area 6A	tibia	45
ER2595	Ileret area 1A	cranial fragments	44
ER2596	Ileret area 15	tibia	49
ER2597	Ileret area 15	M	49
ER2598	Ileret area 15	occipitale	49
ER2599	Ileret area 15	P	49
ER2600	Koobi Fora area 130	M	56
ER2601	Koobi Fora area 130	M	56
ER2602	Koobi Fora area 117	cranial fragments	53
ER2603	Koobi Fora area 117	dens	53
ER2604	Koobi Fora area 117	dens	53
ER2605	Koobi Fora area 117	dens	53

Reg. No.	Area	Element	Page
ER2606	Koobi Fora area 117	dens	53
ER2607	Koobi Fora area 105	dens	52
ER3228	Koobi Fora area 102	os coxae	50
ER3227	Koobi Fora area 103	mandibula	50
ER3230	Koobi Fora area 131	mandibula	56
ER3728	Koobi Fora area 100	femur	50
ER3729	Koobi Fora area 102	mandibula	50
ER3730	Koobi Fora area 102	femur	50
ER3731	Koobi Fora area 105	mandibula	52
ER3732	Koobi Fora area 105	cranium	52
ER3733	Koobi Fora area 104	cranium	51
ER3734	Koobi Fora area 105	mandibula	52
ER3735	Koobi Fora area 116	humerus	53
ER3736	Koobi Fora area 105	radius	52

1. Ileret area 1.
2. On erosion slopes, about 6 km E of Ileret police post.   4° 19′ N, 36° 16′ E.
3. P. Nzube, 1970 (ER725); K. Kimeu, 1970 (ER728); H. Mutua, 1970 (ER739); M. D. Leakey, 1970 (ER741); K. Kimeu, 14 June 1971 (ER805); H. Mutua, 30 August 1971 (ER819, 820); K. Kimeu, 1 November 1971 (ER993); K. Kimeu, 13 June 1972 (ER1463).
4. Ileret Member: below the Lower Tuff (ER819, 820, 1817); between the Middle and Chari Tuffs (ER725, 728, 741, 805, 993, 1463, 1815); stratigraphic position not yet determined (ER739).   B. E. Bowen and C. F. Vondra 1973.
8. *Metridiochoerus andrewsi* zone (ER819, 820).   *Loxodonta africana* zone (remainder except ER739).   V. J. Maglio 1972.
11. **ER725.**   lt corpus mandibulae (f), roots P3–M3.
    ER728.   rt corpus mandibulae (f), roots M1–3.
    ER739.   distal rt humerus (f).
    ER741.   proximal rt tibia (f).
    ER805A. lt corpus mandibulae (f), roots M1–3.
    ER805B. rt corpus mandibulae (f).
    ER819.   lt corpus mandibulae (f), roots P3–M2.
    ER820.   juvenile: corpus mandibulae, lt I1, I2, dc (f), dm1, dm2, M1; rt I1, I2, dm1, dm2, M1.
    ER993.   distal rt femur.
    ER1463. diaphysis rt femur (f).
    ER1815. rt talus (ff).
    ER1817. lt corpus mandibulae (ff).
12. R. E. F. Leakey 1971, Further evidence of Lower Pleistocene hominids from East Rudolf, North Kenya, *Nature, Lond.,* **231** : 241–245 (ER725, 728, 739, 741).   R. E. F. Leakey 1972a (ER805, 819, 820, 993).   R. E. F. Leakey 1973a, Further evidence of Lower Pleistocene hominids from East Rudolf, North Kenya, 1972 *Nature, Lond.,* **242** : 170–173 (ER1463).   *Australopithecus.*   R. E. F. Leakey 1974, Further evidence of Lower Pleistocene hominids from East Rudolf, North Kenya, 1973, *Nature, Lond.,* **248** : 653–656 (ER1815, 1817).
13. R. E. F. Leakey, J. M. Mungai and A. C. Walker 1972, New australopithecines from East Rudolf, Kenya (II), *Am. J. phys. Anthrop.,* **36** : 235–251 (ER725, 728, 739,

741). *Australopithecus*. R. E. F. Leakey and A. C. Walker 1973, New australopithecines from East Rudolf, Kenya (III), *Am. J. phys. Anthrop.*, **39** : 205–221 (ER805A, 819, 993). *Australopithecus*. R. E. F. Leakey and B. A. Wood 1973, New evidence of the genus *Homo* from East Rudolf, Kenya (II), *Am. J. Phys. Anthrop.*, **39** : 355–368 (ER820). *Homo.* M. H. Day, R. E. F. Leakey, A. C. Walker and B. A. Wood 1976, New Hominids from East Turkana, Kenya, *Am. J. phys. Anthrop.*, **45** : 369–436 (ER1463).

14. A. Walker 1973, New *Australopithecus* femora from East Rudolf, Kenya, *J. hum. Evol.*, **2** : 545–555 (ER993).

15. R. E. F. Leakey, J. M. Mungai and A. C. Walker 1972 (ER725, 728, 739, 741). R. E. F. Leakey and B. A. Wood 1973 (ER820). R. E. F. Leakey and A. C. Walker 1973 (ER993). M. H. Day *et al.* 1976 (ER1463).

1. Ileret Area 1A.
4. Ileret Member. R. E. F. Leakey 1976, New fossil hominids from the Koobi Fora formation, Northern Kenya, *Nature, Lond.*, **261** : 574–576.
11. **ER2595**: cranial fragments.
12. R. E. F. Leakey 1976.

1. Ileret area 3.
2. On erosion slopes, about 5 km E of Ileret police post. 4° 18′ N, 36° 16′ E.
3. W. Garland, 1970 (ER740); B. Ngeneo, 1 November 1971 (ER992); P. Nzube, 21 June 1972 (ER1467); 1973 (ER1819).
4. Ileret Member, between the Middle and Chari Tuffs. B. E. Bowen and C. F. Vondra 1973.
8. *Loxodonta africana* zone. V. J. Maglio 1972.
11. **ER740.**   distal diaphysis lt humerus (ff).
     ER992.   corpus mandibulae, lt I2, C–M3, rt C–M3.
     ER1467.   lower lt M3 crown.
     ER1819.   lower lt M3.
12. R. E. F. Leakey 1971, Further evidence of Lower Pleistocene hominids from East Rudolf, North Kenya, *Nature, Lond.*, **231** : 241–245 (ER740). R. E. F. Leakey 1972a (ER992). R. E. F. Leakey 1973a, Further evidence of Lower Pleistocene hominids from East Rudolf, North Kenya, 1972, *Nature, Lond.*, **242** : 170–173 (ER1467). R. E. F. Leakey 1974, Further evidence of Lower Pleistocene hominids from East Rudolf, North Kenya, 1973, *Nature, Lond.*, **248** : 653–656 (ER1819).
13. R. E. F. Leakey, J. M. Mungai and A. C. Walker 1972, New australopithecines from East Rudolf, Kenya (II), *Am. J. phys. Anthrop.*, **36** : 235–251 (ER740). *Australopithecus*. R. E. F. Leakey and B. A. Wood 1973, New evidence of the genus *Homo* remains from East Rudolf, Kenya, (II), *Am. J. phys. Anthrop.*, **39** : 355–368 (ER992). *Homo.* M. H. Day, R. E. F. Leakey, A. C. Walker and B. A.

Wood 1976, New Hominids from East Turkana, Kenya, *Am. J. phys. Anthrop.*, **45**: 369–436 (ER1467).
15.   R. E. F. Leakey and B. A. Wood 1973 (ER992).
16.   R. E. F. Leakey 1973b, Australopithecines and hominines: a summary of the evidence from the early Pleistocene of Eastern Africa, *in* S. Zuckerman (ed.) 1973, The Concepts of Human Evolution, *Symp. Zool. Soc. Lond.*, No. 33 : 53–69 (ER992).

1.   Ileret area 6.
2.   On erosion slope, about 6 km S of Ileret police post.   4° 16′ N, 36° 14′ E.
3.   B. Ngeneo, 14 June 72.
4.   Ileret Member, between the Middle and Chari Tuffs (ER1466).   B. E. Bowen and C. F. Vondra 1973.   Below 'Middle Tuff' (ER2592, ER2593).   R. E. F. Leakey 1976.   New fossil hominids from the Koobi Fora formation Northern Kenya, *Nature, Lond.*, **261** : 574–576.
8.   *Loxodonta africana* zone.   V. J. Maglio 1972.
11.   **ER1466.**   parietale (f).
      ER2592.   parietale (ff).
      ER2593.   M (f).
12.   R. E. F. Leakey 1973a (ER1466).   R. E. F. Leakey, 1976 (ER2592, ER2593).
13.   M. H. Day, R. E. F. Leakey, A. C. Walker and B. A. Wood 1976, New Hominids from East Turkana, Kenya, *Am. J. phys. Anthrop.*, **45**: 369–436 (ER1466).

1.   Ileret area 6A.
2.   On erosion slopes, about 5 km S of Ileret.   4° 17′ N, 36° 14′ E.
3.   H. Mutua, 1970 (ER727); K. Kimeu, 1970 (ER731); R. E. F. Leakey, 6 June 71 (ER801, 802, 1170, 1171); B. Ngeneo, 30 August 1971 (ER818); K. Kimeu, 3 November 1971 (ER999); P. Nzube, 13 June 1972 (ER1464).
4.   Just above the base of the Guomde Formation (ER999); Ileret Member, just below the Chari Tuff (ER731); between the Lower and Middle Tuffs (ER818; below the Lower Tuff (ER727, 801, 802, 1170, 1171, 1464).   B. E. Bowen and C. F. Vondra 1973.   Between Lower and KBS Tuffs (ER1816).   Upper Member (ER1818).   R. E. F. Leakey 1974, Further evidence of Lower Pleistocene hominids from East Rudolf, North Kenya, 1973, *Nature, Lond.*, **248** : 653–656.   Below the 'Middle Tuff' (ER2594 a and b).   R. E. F. Leakey 1976, New hominid fossils from the Koobi Fora formation in Northern Kenya, *Nature, Lond.*, **261** : 574–576.
8.   *Loxodonta africana* zone (ER731, 818); associated fauna unknown (ER999); *Metridiochoerus andrewsi* zone (remainder).   V. J. Maglio 1972.
11.   **ER727.**   rt corpus mandibulae (f), roots M1–M2.
      ER731.   lt corpus mandibulae (f), alveoli C–M1.
      ER801.   rt corpus mandibulae (f), M2, M3, roots M1: other dentes (ff).
      ER802.   upper rt M1, lower rt P4, M1 (f), M3; upper lt C (f), lower lt P4, M1 (f), M2 (f), M3 (f).

ER818.   lt corpus mandibulae (f), P3 (ff), P4, M1 (f), M2 (f), M3 (f), I2–C roots.
ER999.   lt femur lacking lateral distal condyle.
ER1170.   cranial fragments.
ER1171A–I.   juvenile: germs upper rt C (f), M2, lt M2; lower lt I2, germs lower lt P4,
                  M2 or 3, rt M1 or 2 (ff), M2 or 1.
ER1464.
ER1816.   juvenile: corpus mandibulae (f).
ER1818.   upper lt I1.
ER1823.   metatarsale (proximal part).
ER1824.   humerus (distal).
ER1825.   atlas (f).
ER2594 a & b. tibia proximal fragment; tibia diaphysis (f).

12. R. E. F. Leakey 1971, Further evidence of Lower Pleistocene hominids from East Rudolf, North Kenya, *Nature, Lond.,* **231** : 241–245 (ER727, 731).   R. E. F. Leakey 1973a, Further evidence of Lower Pleistocene hominids from East Rudolf, North Kenya, 1972, *Nature, Lond.,* **242** : 170–173 (ER1464).   M. H. Day and R. E. F. Leakey 1974, New evidence of the Genus *Homo* from East Rudolf, Kenya (III), *Am. J. phys. Anthrop.,* **41** : 367–380 (ER999).   R. E. F. Leakey 1972a (ER801, 802, 818, 1170, 1171).   R. E. F. Leakey 1974 (ER1816, 1818).   R. E. F. Leakey, 1976 (ER2594a & b).

13. R. E. F. Leakey, J. M. Mungai and A. C. Walker 1972, New australopithecines from East Rudolf, Kenya (II), *Am. J. phys. Anthrop.,* **36** : 235–251 (ER727).   *Australopithecus.*   R. E. F. Leakey and A. C. Walker 1973, New Australopithecines from East Rudolf, Kenya (III), *Am. J. phys. Anthrop.,* **39** : 205–221 (ER801, 802, 818, 1171).   *Australopithecus.*   M. H. Day and R. E. F. Leakey 1973, New evidence of the Genus *Homo* from East Rudolf, Kenya (I), *Am. J. phys. Anthrop.,* **39** : 341–354 (ER731).   *Homo.*   M. H. Day, R. E. F. Leakey, A. C. Walker and B. A. Wood 1976, New Hominids from East Turkana, Kenya, *Am. J. phys. Anthrop.,* **45** : 369–436 (ER1816, 1464, 1823).

15. R. E. F. Leakey, J. M. Mungai and A. C. Walker 1972 (ER727).   R. E. F. Leakey and A. C. Walker 1973 (ER801, 818).   M. H. Day *et al.* 1976 (ER1464).

1. Ileret area 7A.
2. On erosion slope, about 10 km SE of Ileret police post.   4° 14′ N, 36° 16′ E.
3. K. Kimeu, 1968.
4. Ileret Member, just below the Chari Tuff.   B. E. Bowen and C. F. Vondra 1973.
8. *Loxodonta africana* zone.   V. J. Maglio 1972.
11. **ER404.**   old adult: rt corpus mandibulae, M2 (ff), M3 (ff).
12. R. E. F. Leakey 1970, Fauna and artefacts from a new Plio-Pleistocene locality near Lake Rudolf in Kenya, *Nature, Lond.,* **226** : 223–224.
13. R. E. F. Leakey, J. M. Mungai and A. C. Walker 1971, New australopithecines from East Rudolf, Kenya, *Am. J. phys. Anthrop.,* **35** : 175–186.
15. R. E. F. Leakey, J. M. Mungai and A. C. Walker 1971.

1. Ileret area 8.
2. On erosion slopes, about 9 km SE of Ileret police post.   4° 16′ N, 36° 18′ E.
3. P. Abell, 1970 (ER729); B. Ngeneo, 22 June 1972 (ER1468).
4. Ileret Member, just above the Lower Tuff (ER729); between the Lower and Middle Tuffs (ER1468).   B. E. Bowen and C. F. Vondra 1973.
8. *Loxodonta africana* zone.   V. J. Maglio 1972.
11. **ER729.**   corpus mandibulae, lt I1 (f), C, P3, P4 (f), M2 (f), M3; rt I1 (f), C (ff), P4 (f), M1 (f), M2 (f), M3.
     ER1468.   rt corpus mandibulae, roots P4–M3.
12. R. E. F. Leakey 1971, Further evidence of Lower Pleistocene hominids from East Rudolf, North Kenya, *Nature, Lond.,* **231** : 241–245 (ER729).   R. E. F. Leakey 1973a, Further evidence of Lower Pleistocene hominids from Eastern Rudolf, North Kenya, 1972, *Nature, Lond.,* **242** : 170–173 (ER1468).
13. R. E. F. Leakey, J. M. Mungai and A. C. Walker 1972, New australopithecines from East Rudolf, Kenya (II), *Am. J. phys. Anthrop.,* **36** : 235–251 (ER729).   *Australopithecus.*   M. H. Day, R. E. F. Leakey, A. C. Walker and B. A. Wood 1976, New Hominids from East Turkana, Kenya, *Am. J. phys. Anthrop.,* **45**: 369–436 (ER1468).
15. R. E. F. Leakey, J. M. Mungai and A. C. Walker 1972 (ER729).   M. H. Day *et al.* 1976 (ER1468).

1. Ileret area 8A.
2. On erosion slopes, about 10 km SE of Ileret police post.   4° 15′ N, 36° 17′ E.
3. R. E. F. Leakey, 1970 (ER733); M. G. Leakey, 8 June 1971 (ER803); H. Mutua, 17 June 1971 (ER806); P. Nzube, 19 June 1971 (ER807); P. Nzube, 26 June 1971 (ER808); K. Kimeu, 26 June 1971 (ER809A).
4. Ileret Member, just above the Lower Tuff (ER806); between the Lower and Middle Tuffs (remainder).   B. E. Bowen and C. F. Vondra 1973.
8. *Loxodonta africana* zone.   V. J. Maglio 1972.
11. **ER733.**   cranial fragments; lt maxilla (f), M1 roots, P3, P4; rt corpus mandibulae (f), M3.
     ER803.   upper rt I1, lt C, diaphysis rt radius (f), diaphysis lt ulna (f), diaphysis lt femur, diaphysis lt tibia, diaphyses 2 fibulae (f), lt talus (f), lt metatarsale III (f), proximal lt metatarsale V, 4 phalanges, post-cranial fragments.
     ER806.   lower lt P3, M1–M3; lower rt P3, M3.
     ER807.   rt maxilla (f), M2 (f), M3.
     ER808A.   juvenile: upper lt & rt dil, rt M1 (f), germs P3, lt & rt I1, rt I2 (f).
     ER809A.   lower lt M1, lower M (ff), dentes (f).
12. R. E. F. Leakey 1971, Further evidence of Lower Pleistocene hominids from East Rudolf, North Kenya, *Nature, Lond.,* **231** : 241–245 (ER733).   R. E. F. Leakey 1972a (remainder).
13. R. E. F. Leakey and A. C. Walker 1973, New australopithecines from East Rudolf, Kenya (III), *Am. J. phys. Anthrop.,* **39** : 205–221 (ER733) *Australopithecus.*   R. E. F. Leakey and B. A. Wood 1973, New *Homo* remains from East Rudolf, Kenya (II), *Am.*

J. phys. Anthrop., **39** : 355–368 (ER806–9A) *Homo.*   M. H. Day and R. E. F. Leakey 1974, New Evidence of the Genus *Homo* from East Rudolf, Kenya (III), *Am. J. phys. Anthrop.*, **41** : 367–380.

15.   R. E. F. Leakey and A. C. Walker 1973 (ER733).   R. E. F. Leakey 1972a (ER803).

1.   Ileret area 10.

2.   On erosion slopes, about 11 km E of Ileret police post.   4° 18′ N, 36° 19′ E.

3.   R. E. F. Leakey and M. G. Leakey, 1969 (ER406); M. Muoka, 1969 (ER407); H. Mutua, 1970 (ER726, 732); A. Hill, 1971 (ER815).

4.   Ileret Member, being from well above the KBS Tuff (ER406, 732); between the KBS and Lower Tuffs (ER815); provenance to be confirmed (ER407, 726).   B. E. Bowen and C. F. Vondra 1973.

8.   *Metridiochoerus andrewsi* zone (excepting ER407, 726).   V. J. Maglio 1972.

11.   **ER406.**   cranium, upper dentes roots.
    ER407.   calvaria (f).
    ER726.   lt corpus mandibulae (f), roots P3–M3.
    ER732.   rt half of cranium (f), P4 (f), roots M1–3.
    ER815.   proximal diaphysis and neck lt femur.

12.   R. E. F. Leakey 1970, Fauna and artefacts from a new Plio-Pleistocene locality near Lake Rudolf in Kenya, *Nature, Lond.*, **226** : 223–224 (ER406, 407).   R. E. F. Leakey 1971, Further evidence of Lower Pleistocene hominids from East Rudolf, North Kenya, *Nature, Lond.*, **231** : 241–245 (ER726, 732).   R. E. F. Leakey 1972a, Further evidence of Lower Pleistocene hominids from East Rudolf, North Kenya, 1971, *Nature, Lond.*, **237** : 264–269 (ER815).

13.   R. E. F. Leakey, J. M. Mungai and A. C. Walker, 1971, New australopithecines from East Rudolf, Kenya, *Am. J. phys. Anthrop.*, **35** : 175–186 (ER406, 407).   *Australopithecus.*   R. E. F. Leakey, J. M. Mungai and A. C. Walker 1972, New australopithecines from East Rudolf, Kenya (II), *Am. J. phys. Anthrop.*, **36** : 235–251 (ER726, 732).   *Australopithecus.*   R. E. F. Leakey and A. C. Walker 1973, New australopithecines from East Rudolf, Kenya (III), *Am. J. phys. Anthrop.*, **39** : 205–221 (ER815).   *Australopithecus.*

14.   A. Walker 1973, New *Australopithecus* femora from East Rudolf, Kenya, *J. hum. Evol.*, **2** : 545–555 (ER815).   R. E. F. Leakey and A. C. Walker 1976, *Australopithecus*, *Homo erectus* and the single species hypothesis, *Nature, Lond.*, **261** : 572–574 (ER406).   M. H. Day, R. E. F. Leakey, A. C. Walker and B. A. Wood 1976, New Hominids from East Turkana, Kenya, *Am. J. phys. Anthrop.*, **45**: 369–436 (ER407).

15.   R. E. F. Leakey, J. M. Mungai and A. C. Walker 1971 (ER406, 407).   R. E. F. Leakey, J. M. Mungai and A. C. Walker 1972 (ER726, 732).   R. E. F. Leakey and A. C. Walker 1973 (ER815).   M. H. Day *et al.* 1976 (ER407).

16.   R. E. F. Leakey 1973c, Australopithecines and hominines: a summary of the evidence from the early Pleistocene of Eastern Africa, *in* S. Zuckerman (ed.), 1973, The Concepts of Human Evolution, *Symp. Zool. Soc. Lond.*, No. 33 : 53–69.   R. L. Holloway 1973, Endocranial Volumes of Early African Hominids, and the Role of the Brain in Human Mosaic Evolution, *J. hum. Evol.*, **2** : 449–459 (ER406, 732).   R. E. F. Leakey 1976, Hominids in Africa, *Am. Sci.*, **64** : 174–178 (ER406).

1. Ileret area 11.
2. On erosion slope, about 8 km E of Ileret police post.   4° 18′ N, 36° 18′ E.
3. K. Kimeu, 13 June 1972.
4. Ileret Member, between the Middle and Chari Tuffs.   B. E. Bowen and C. F. Vondra 1972.
8. *Loxodonta africana* zone.   V. J. Maglio 1972.
11. **ER1465.**   proximal lt femur, lacking head.
12. R. E. F. Leakey 1973a, Further evidence of Lower Pleistocene hominids from East Rudolf, North Kenya, 1972, *Nature, Lond.,* **242** : 170–173.
13. M. H. Day, R. E. F. Leakey, A. C. Walker and B. A. Wood 1976, New Hominids from East Turkana, Kenya, *Am. J. phys. Anthrop.,* **45**: 369–436.

1. Ileret area 12.
2. On erosion slopes, about 11 km SE of Ileret police post.   4° 16′ N, 36° 18′ E.
3. B. Ngeneo, 18 November 1972 (ER1590); H. Mutua, 18 November 1972 (ER1591); P. Nzube, 18 November 1972 (ER1592); M. Muliula, 25 November 1972 (ER1593).
4. Lower member, just below the KBS Tuff (ER1590, 1593); Ileret Member, between the KBS and Lower Tuffs (ER1591, 1592).   B. E. Bowen and C. F. Vondra 1973.
8. *Mesochoerus limnetes* zone (ER1590, 1593); *Metridiochoerus andrewsi* zone (ER1591, 1592).   V. J. Maglio 1972.
11. **ER1590.**   juvenile: cranial fragments, dc, dm2, upper lt and rt C–M1.
    ER1591.   humerus lacking head.
    ER1592.   distal half femur.
    ER1593.   cranial and mandibular fragments.
12. R. E. F. Leakey 1973a Further evidence of Lower Pleistocene hominids from East Rudolf, North Kenya, 1972, *Nature, Lond.,* **242** : 170–173.   (ER1590, 1591, 1593) *Homo;* (ER1592) *Australopithecus.*
13. R. E. F. Leakey 1974, Further evidence of Lower Pleistocene hominids from East Rudolf, North Kenya, 1973, *Nature, Lond.,* **248** : 653–656 (ER1590).   *Homo.*   M. H. Day, R. E. F. Leakey, A. C. Walker and B. A. Wood 1976, New Hominids from East Turkana, Kenya, *Am. J. phys. Anthrop.,* **45**: 369–436 (ER1590, 1593).
14. ─────────
15. M. H. Day *et al.* 1976 (ER1590).

1. Ileret Area 15.
4. Upper Member (ER2596).   Lower Member (ER2597, ER2598, ER2599).   R. E. F. Leakey 1976, New fossil hominids from the Koobi Fora formation, Northern Kenya, *Nature, Lond.,* **261** : 574–576.
11. **ER2596.**   distal fragt tibia.
    ER2597.   M crown.
    ER2598.   occipitale (ff).
    ER2599.   P crown (ff).
12. R. E. F. Leakey 1976.

1. Koobi Fora Area 100.
3. M. Muluila 1974–5.
4. ?Lower Member.   R. E. F. Leakey 1976, New fossil hominids from the Koobi Fora formation in Northern Kenya, *Nature, Lond.,* **261** : 574–576.
11. **ER3728.** lt femur.
12. R. E. F. Leakey 1976. *Australopithecus.*

1. Koobi Fora Area 102.
3. B. Ngeneo 1974–5 (ER3228).
4. Lower Member, eroded from a fluvial lens within a generally lacustrine section (ER3228). R. E. F. Leakey 1976, New hominids from the Koobi Fora formation in Northern Kenya, *Nature, Lond.,* **261** : 574–576.
11. **ER3228.** adult male: right os coxae.
    ER3729. lt corpus mandibulae, C–M3 roots (f).
    ER3730. distal femur.
12. R. E. F. Leakey 1976 (ER3228). *Homo.*
13. ————
14. ————
15. R. E. F. Leakey 1976.

1. Koobi Fora area 103.
2. On erosion slopes, about 46 km S of Ileret police post.   3° 54′ N, 36° 14′ E.
3. M. Muoka, 1968 (ER403); M. G. Leakey, 1970 (ER730, 736); R. E. F. Leakey, 1970 (ER734, 737); P. Nzube, 14 October 1972 (ER1515); K. Kimeu, 1973 (ER1807, 1808, 1820).
4. Upper Member, immediately below the Koobi Fora Tuff.   B. E. Bowen and C. F. Vondra 1973.
8. *Metridiochoerus andrewsi* zone.   V. J. Maglio 1972.
11. **ER403.**   rt corpus mandibulae (f), P3–M3 (f), roots I1, I2, C.
    ER730.   lt corpus mandibulae (f) and symphysial region, lt M1 (f), M2 (f), M3, roots rt I2, lt and rt C, P3–4.
    ER734.   parietal (f).
    ER736.   diaphysis lt femur.
    ER737.   diaphysis lt femur.
    ER1515.   upper rt I1.
    ER1807.   femur diaphysis (f).
    ER1808.   skull (ff), upper rt C, M1 or 2, lower rt M2 or 3, roots of lower rt M2 and M3; rt humerus (ff), rt femur.
    ER1820.   juvenile: lt corpus mandibulae, M1, I2 (unerupted), roots of dc, dm1, dm2.
    ER3227.   corpus mandibulae, rt and lt P3.
12. R. E. F. Leakey 1970, Fauna and artefacts from a new Plio-Pleistocene locality near

Lake Rudolf in Kenya, *Nature, Lond.,* **226** : 223–224 (ER403).   R. E. F. Leakey 1973a, Further evidence of Lower Pleistocene hominids from East Rudolf, North Kenya, 1972, *Nature, Lond.,* **242** : 170–173 (ER1515).   R. E. F. Leakey 1971, Further evidence of Lower Pleistocene hominids from East Rudolf, North Kenya, *Nature, Lond.,* **231** : 241–245 (ER730, 734, 736, 737).   R. E. F. Leakey 1974, Further evidence of Lower Pleistocene hominids from East Rudolf, North Kenya, 1973, *Nature, Lond.,* **248** : 653–656 (ER1807, 1808, 1820).   R. E. F. Leakey 1976, New hominid fossils from the Koobi Fora formation in Northern Kenya, *Nature, Lond.,* **261** : 574–576 (ER3227).

13.   R. E. F. Leakey, J. M. Mungai and A. C. Walker 1971, New australopithecines from East Rudolf, Kenya, *Am. J. phys. Anthrop.,* **35** : 175–186 (ER403).   *Australopithecus.*   R. E. F. Leakey, J. M. Mungai and A. C. Walker 1972, New Australopithecines from East Rudolf, Kenya, (II), *Am. J. phys. Anthrop.,* **36** : 235–251 (ER736).   *Australopithecus.*   M. H. Day and R. E. F. Leakey 1973, New evidence of the Genus *Homo* from East Rudolf, Kenya, *Am. J. phys. Anthrop.,* **39** : 341–354 (ER737).   *Homo.*   M. H. Day, R. E. F. Leakey, A. C. Walker and B. A. Wood 1976, New Hominids from East Turkana, Kenya, *Am. J. phys. Anthrop.,* **45**: 369–436 (ER1820, 1807).

14.   A. Walker 1973, New *Australopithecus* Femora from East Rudolf, Kenya, *J. hum. Evol.,* **2** : 545–555 (ER736).

15.   R. E. F. Leakey, J. M. Mungai and A. C. Walker 1971 (ER403).   R. E. F. Leakey, J. M. Mungai and A. C. Walker 1972 (ER736).   M. H. Day and R. E. F. Leakey 1973 (ER730, 737).   M. H. Day *et al.* 1976 (ER1820).

1.   Koobi Fora area 104.
2.   On erosion slopes, about 38 km SSE of Ileret police post.   3° 59′ N, 36° 19′ E.
3.   H. Stradling, 1969 and M. G. Leakey, 7 June 1971 (ER164); Kitivi Kimeu, 1 July 1971 (ER810, 998); B. Ngeneo, 2 July 1971 (ER811); B. Ngeneo, 1 July 1971 (ER812, 813); B. Ngeneo, 12 July 1971 (ER814); P. Nzube, 12 July 1971 (ER816, 997).   R. Holloway, 1973 (ER1804).   B. Ngeneo, 1974–5 (ER3733).
4.   Upper member, near the level of the white tuff (ER164); stratigraphic position not determined (ER811).   B. E. Bowen and C. F. Vondra 1973.
8.   *Metridiochoerus andrewsi* zone (except ER811).   V. J. Maglio 1972.
11.   **ER164.**   parietale (f), 2 vertebrae cervicales ? (f), 2 phalanges manus (f).
     ER810.   lt corpus mandibulae (f), M3.
     ER811.   parietale (f).
     ER812.   juvenile: lt corpus mandibulae (f) and symphysial region, germs rt I1 (ff), lt I1, C–M1 in crypts, roots lt dc–dm2.
     ER813.   (A) rt talus (f); (B) shaft rt tibia (f).
     ER814.   cranial fragments.
     ER816.   C, M (f).
     ER997.   lt metatarsale III.
     ER998.   I.
     ER1804.   lt maxilla (ff), C (root), P3–M2.
     ER3733.   cranium.

12. R. E. F. Leakey 1972a, Further evidence of Lower Pleistocene hominids from East Rudolf, North Kenya, 1971, *Nature, Lond.,* **237** : 264–269.   R. E. F. Leakey 1974, Further evidence of Lower Pleistocene hominids in East Rudolf, North Kenya, 1973, *Nature, Lond.,* **248** : 653–656 (ER1804).   R. E. F. Leakey 1976, New fossil hominids from the Koobi Fora formation in Northern Kenya, *Nature, Lond.,* **261** : 574–576 (ER3733).

13. R. E. F. Leakey and A. C. Walker 1973, New australopithecines from East Rudolf, Kenya (III), *Am. J. phys. Anthrop.,* **39** : 205–221 (ER810, 812, 814).   *Australopithecus.*   R. E. F. Leakey and B. A. Wood 1973, New evidence of the Genus *Homo* from East Rudolf, Kenya (II), *Am. J. phys. Anthrop.,* **39** : 355–368 (ER813).   *Homo.*   M. H. Day and R. E. F. Leakey 1974, New Evidence of the Genus *Homo* from East Rudolf, Kenya (III), *Am. J. phys. Anthrop.,* **41** : 367–380 (ER164).   R. E. F. Leakey and A. C. Walker 1976, *Australopithecus, Homo erectus* and the single species hypothesis, *Nature, Lond.,* **261** : 572–574 (ER3733).   *Homo erectus.*   M. H. Day, R. E. F. Leakey, A. C. Walker and B. A. Wood 1976, New Hominids from East Turkana, Kenya, *Am. J. phys. Anthrop.,* **45**: 369–436 (ER1804).

15. R. E. F. Leakey and A. C. Walker 1973 (ER810, 812).   R. E. F. Leakey and B. A. Wood 1973 (ER813).   M. H. Day *et al.* 1976 (ER1804).

1. Koobi Fora area 105.
2. On erosion slopes, about 32 km SE of Ileret police post.   4° 04′ N, 36° 22′ E.
3. P. Nzube, 1968 (ER405); B. Ngeneo, 1971 (ER738); K. Kimeu, 29 July 1972 (ER1476); M. Mbithi, 29 July 1972 (ER1477); R. E. F. Leakey, 9 August 1972 (ER1478, 1479); J. M. Harris and A. K. Behrensmeyer, 1972 (ER1480); M. Mangoka, 2 August 1971 (ER1648).   1974–5 (ER3731, ER3732, ER3734).
4. Upper member; just above the disconformity above the KBS Tuff (ER405); above the KBS Tuff (ER1480, ER2607, ER3736); stratigraphic position not determined (ER1648); from or above the KBS Tuff (remainder).   B. E. Bowen and C. F. Vondra 1973.   Lower member (ER3734).   R. E. F. Leakey 1976, New fossil hominids from the Koobi Fora formation Northern Kenya, *Nature, Lond.,* **261** : 574–576.
7. Artifacts recovered from the KBS Tuff in this area but not in direct association with hominid remains.   M. D. Leakey 1970.   G. Ll. Isaac, R. E. F. Leakey and A. K. Behrensmeyer 1971.
8. *Metridiochoerus andrewsi* zone (except ER1648).   V. J. Maglio 1972.
11. **ER405.**   palatinum, dentes (roots).
      ER738.   proximal half lt femur.
      ER1476.   lt talus (f), proximal lt tibia, diaphysis rt tibia.
      ER1477.   juvenile: corpus mandibulae (f), lt and rt dc–dm2, germs lt I2–P3, M1 and rt M1.
      ER1478.   cranial fragments.
      ER1479.   lower lt and rt M3 (crowns).
      ER1480.   lower rt M3 (crown).
      ER1648.   cranial fragment.
      ER2607.   dens (ff).
      ER3731.   lt corpus mandibulae, 1–M3 roots (f).

> **ER3732.**   cranium (ff).
> **ER3734.**   lt corpus mandibulae, 1–M2, M3 (f).
> **ER3736.**   proximal radius.

12.   R. E. F. Leakey 1970, Fauna and artefacts from a new Plio-Pleistocene locality near Lake Rudolf in Kenya, *Nature, Lond.,* **226** : 223–224 (ER405).   R. E. F. Leakey 1971, Further evidence of Lower Pleistocene hominids from East Rudolf, North Kenya, *Nature, Lond.,* **231** : 241–245 (ER738).   Unpublished (ER1648).   R. E. F. Leakey 1973a, Further evidence of Lower Pleistocene hominids from East Rudolf, North Kenya, 1972, *Nature, Lond.,* **242** : 170–173 (remainder).   R. E. F. Leakey 1976, (ER2607, ER3731, ER3732, ER3734, ER3736).

13.   R. E. F. Leakey, J. M. Mungai and A. C. Walker 1971, New australopithecines from East Rudolf, Kenya, *Am. J. phys. Anthrop.,* **35** : 175–186 (ER405). *Australopithecus.*   R. E. F. Leakey, J. M. Mungai and A. C. Walker 1972, New Australopithecines from East Rudolf, Kenya (II), *Am. J. phys. Anthrop.,* **36** : 235–251 (ER738). *Australopithecus.*   R. E. F. Leakey and B. A. Wood 1974, New Evidence of the Genus *Homo* from East Rudolf, Kenya (IV), *Am. J. phys. Anthrop.,* **41** : 237–243 (ER1480).   M. H. Day, R. E. F. Leakey, A. C. Walker and B. A. Wood 1976, New Hominids from East Turkana, Kenya, *Am. J. phys. Anthrop.,* **45**: 369–436 (ER1477, 1478, 1476).

14.   A. Walker 1973, New *Australopithecus* Femora from East Rudolf, Kenya, *J. hum. Evol.,* **2** : 545–555 (ER738).

15.   R. E. F. Leakey, J. M. Mungai and A. C. Walker 1971 (ER405).   R. E. F. Leakey, J. M. Mungai and A. C. Walker 1972 (ER738).   R. E. F. Leakey 1973a, Further evidence of Lower Pleistocene hominids from East Rudolf, North Kenya, 1972, *Nature, Lond.,* **242** : 170–173 (ER1477).

 1.   Koobi Fora Area 116.
 3.   1974–5.
 4.   Lower Member.   R. E. F. Leakey 1976, New fossil hominids from the Koobi Fora formation, Northern Kenya, *Nature, Lond.,* **261** : 574–576.
11.   **ER3735.**   rt distal humerus (f).
12.   R. E. F. Leakey 1976.

 1.   Koobi Fora Area 117.
 4.   Lower Member.   R. E. F. Leakey 1976, New fossil hominids from the Koobi Fora formation, Northern Kenya, *Nature, Lond.,* **261** : 574–576.
11.   **ER2602.**   cranial fragments.
     ER2603.   dens (ff).
     ER2604.   dens (ff).
     ER2605.   dens (ff).
     ER2606.   dens (ff).
12.   R. E. F. Leakey 1976.

 1.   Koobi Fora area 119.
 2.   On erosion slope, about 46 km SSE of Ileret police post.   3° 54′ N, 36° 18′ E.
 3.   H. Mutua, 1972 (ER1509); P. Nzube, 1972 (ER1510).
 4.   Not yet determined.
11.   **ER1509.**   lower lt M1, M2, crowns lt C, P3 or P4 & M3.
      ER1510.
12.   R. E. F. Leakey 1973a, Further evidence of Lower Pleistocene hominids from East Rudolf, North Kenya, 1972, *Nature, Lond.*, **242** : 170–173 (ER 1509). *Australopithecus;* (ER 1510). *Homo.*

 1.   Koobi Fora area 121.
 2.   On erosion slope, about 48 km SSE of Ileret police post.   3° 35′ N, 36° 23′ E.
 3.   B. Ngeneo, 1972 (ER1506).
 4.   Just below KBS Tuff.
11.   **ER1506.**   rt corpus mandibulae (f), M1 and M2, isolated crowns lt and rt P3.
12.   R. E. F. Leakey 1973a, Further evidence of Lower Pleistocene hominids from East Rudolf, North Kenya, 1972, *Nature, Lond.*, **242** : 170–173.   *Australopithecus.*
13.   M. H. Day, R. E. F. Leakey, A. C. Walker and B. A. Wood 1976, New hominids from East Turkana, Kenya, *Am. J. phys. Anthrop.*, **45** : 369–436.
14.   ———————
15.   M. H. Day *et al.* 1976.

 1.   Koobi Fora area 123.
 2.   On erosion slopes, about 52 km SSE of Ileret police post.   3° 52′ N, 36° 21′ E.
 3.   B. Ngeneo, 30 November 1972 (ER1501, 1505); W. Cheba, 30 November 1972 (ER1502); M. Muluila, 30 November 1972 (ER1503); M. Mbithi, 30 November 1972 (ER1504); K. Kimeu, 1973 (ER1810, 1811, 1812, 1813).
 4.   Just above KBS Tuff.
11.   **ER1501.**   rt corpus mandibulae (f), roots C–M3.
      ER1502.   rt corpus mandibulae (f), M1 (f), roots M2.
      ER1503.   proximal rt femur.
      ER1504.   distal rt humerus.
      ER1505.   head and fragment distal diaphysis lt femur.
      ER1810.   proximal end of tibia.
      ER1811.   lt corpus mandibulae (ff), roots of I2–P4.
      ER1812.   corpus mandibulae (f), roots of rt I1, I2, C, P3 (f); lt I1, M3; radius (head).
      ER1813.   cranium, upper lt P3–M3, rt I1–C, roots of P3–M2.
      ER1821.   rt parietale including bregma.
      ER1822.   femur diaphysis.
12.   R. E. F. Leakey 1973a, Further evidence of Lower Pleistocene hominids from East Rudolf, North Kenya, 1972, *Nature, Lond.*, **242** : 170–173 (ER1501, 1502). *Homo;* (ER1503, 1504, 1505). *Australopithecus.*   R. E. F. Leakey 1974, Further evidence of

Lower Pleistocene hominids from East Rudolf, North Kenya, 1973, *Nature, Lond.,* **248** : 653–656 (ER1810, 1811, 1812, 1813).

13.　R. E. F. Leakey and B. A. Wood 1974, New evidence of the Genus *Homo* from East Rudolf, Kenya (IV), *Am. J. phys. Anthrop.,* **41** : 237–243 (ER1501, 1502).　M. H. Day, R. E. F. Leakey, A. C. Walker and B. A. Wood 1976, New Hominids from East Turkana, Kenya, *Am. J. phys. Anthrop.,* **45**: 369–436 (ER1811, 1812, 1813, 1503, 1504, 1505, 1810, 1812, 1822).

15.　R. E. F. Leakey 1973a, R. E. F. Leakey 1974 (ER1813).　R. E. F. Leakey and B. A. Wood 1974 (ER1502).　M. H. Day *et al.* 1976 (ER1813, 1503, 1504, 1510).

16.　R. E. F. Leakey 1976, Hominids in Africa, *Am. Sci.,* **64** : 174–178 (ER1813).

1.　Koobi Fora area 124.
2.　On erosion slope, about 45 km SSE of Ileret police post.　3° 56′ N, 36° 21′ E.
3.　P. Nzube, 28 July 1971.
4.　Not yet determined.
11.　**ER817.**　lt corpus mandibulae (f), roots P4.
12.　R. E. F. Leakey 1972a, Further evidence of Lower Pleistocene hominids from East Rudolf, North Kenya, 1971, *Nature, Lond.,* **237** : 264–269.
13.　R. E. F. Leakey and B. A. Wood 1973, New evidence of the genus *Homo* from East Rudolf, Kenya (II), *Am. J. phys. Anthrop.,* **39** : 355–368.　*Homo.*

1.　Koobi Fora area 127.
2.　On erosion slopes, about 46 km SSE of Ileret police post.　3° 56′ N, 36° 24′ E.
3.　P. Nzube, 1972 (ER1507, 1508); 1973 (ER1809, 1814).
11.　**ER1507.**　juvenile: lt corpus mandibulae (f), dm1 (i), M1 (i), germs I, M2.
　　ER1508.　lower rt M2.
　　ER1809.　rt femur diaphysis.
　　ER1814.　maxilla, upper lt P3–M1.
12.　R. E. F. Leakey 1973a, Further evidence of Lower Pleistocene hominids from East Rudolf, North Kenya, 1972, *Nature, Lond.,* **242** : 170–173 *Homo.*　R. E. F. Leakey 1974, Further evidence of Lower Pleistocene hominids from East Rudolf, North Kenya, 1973, *Nature, Lond.,* **248** : 653–656 (ER1809, 1814).
13.　R. E. F. Leakey and B. A. Wood 1974, New evidence of the Genus *Homo* from East Rudolf, Kenya (IV), *Am. J. phys. Anthrop.,* **41** : 237–243 (ER1507, 1508).　M. H. Day, R. E. F. Leakey, A. C. Walker and B. A. Wood 1976, New Hominids from East Turkana, Kenya, *Am. J. phys. Anthrop.,* **45**: 369–436 (ER1809).
15.　R. E. F. Leakey 1973a, (ER1507).

1. Koobi Fora area 129 and 130.
2. On erosion slopes, area 129 (ER417), area 130 (rest) about 26 km SE of Ileret police post.   4° 11′ N, 36° 25′ E.
3. P. Nzube, 1968 (ER417); P. Nzube, 3 July 1972 (ER1462); J. Kimene, 15 November 1972 (ER1500); 1973 (ER1800); P. Abell, 1973 (ER1805); M. Leakey 1973 (ER1806).
4. Lower member below the KBS Tuff (ER417, 1462, 1500, 1800). *In situ* or just below Okote Tuff (ER1805, 1806). B. E. Bowen and C. F. Vondra 1973. Below KBS Tuff (ER2600, ER2601). R. E. F. Leakey 1976. New fossil hominids from the Koobi Fora formation Northern Kenya, *Nature, Lond.,* **261** : 574–576.
8. *Mesochoerus limnetes* zone. V. J. Maglio 1972.
11. **ER417.**   parietale (f).
    ER1462.   lower lt M3.
    ER1500.   rt ulna, rt radius, lt proximal diaphysis ulna (f), proximal and distal femora (ff), lt tibia (ff), rt distal tibia and fibula, rt proximal metatarsale III.
    ER1800.   cranial fragments from rt side.
    ER1805.   cranium (ff), mandibula, upper lt I2–M3, rt I1–M3; lower rt M2, M3.
    ER1806.   corpus mandibulae (f), roots of lt C–M2, rt P3 (f), M2.
    ER2600.   M crown (f).
    ER2601.   M.
12. R. E. F. Leakey 1971, Further evidence of Lower Pleistocene hominids from East Rudolf, North Kenya, *Nature, Lond.,* **231** : 241–245 (ER417). R. E. F. Leakey 1973a, Further evidence of Lower Pleistocene hominids from East Rudolf, North Kenya, 1972, *Nature, Lond.,* **242** : 170–173 (ER1500). *Australopithecus;* (ER1462). *Homo.* R. E. F. Leakey 1974, Further evidence of Lower Pleistocene hominids from East Rudolf, North Kenya, 1973, *Nature, Lond.,* **248** : 653–656 (ER1800, 1805, 1806). R. E. F. Leakey 1976 (ER2600, ER2601).
13. R. E. F. Leakey and B. A. Wood 1974, New evidence of the Genus *Homo* from East Rudolf, Kenya (IV), *Am. J. phys. Anthrop.,* **41** : 237–243 (ER1462). M. H. Day, R. E. F. Leakey, A. C. Walker and B. A. Wood 1976, New Hominids from East Turkana, Kenya, *Am. J. phys. Anthrop.,* **45**: 369–436 (ER1805, 1806, 1500).
14. ————
15. M. H. Day *et al.* 1976 (ER1805, 1500).

1. Koobi Fora area 131.
2. On erosion slopes, about 27 km SE of Ileret police post.   4° 9′ N, 36° 24′ E.
3. K. Kimeu, 3 July 1972 (ER1469); B. Ngeneo, 29 July 1972 (ER1470, 1471, 1473, 1474); J. M. Harris, 29 July 1972 (ER1472); K. Kimeu, 29 July 1972 (ER1475); J. M. Harris, 12 August 1972 (ER1481); M. Muluila, 22 August 1972 (ER1482); W. Cheba, 22 August 1972 (ER1483) 1973 (ER1801); J. Harris, 1973 (ER1802); 1973 (ER1803); Kitibi Kimeu, 1974–5 (ER3230).
4. Lower member at the following depths below the KBS Tuff: 3·5 m (ER1483), 11·25 m (ER1469), 11·75 m (ER1481), 24 m (ER1475), 24·5 m (ER1473), 27 m (ER1471), 30 m (ER1472), 31·5 m (ER1474), 33·75 m (ER1482) and 35·5 m (ER1470). B. E. Bowen and C. F. Vondra 1973. Archaeological site FxJi 20 (east), below Karari Tuff (ER3230). R. E. F. Leakey 1976, New hominid fossils from the Koobi Fora for-

mation Northern Kenya, *Nature, Lond.,* **261** : 574–576.   J. W. K. Harris and G. Isaac 1976, The Karari Industry: Early Pleistocene archaeological evidence from the terrain east of Lake Turkana, Kenya, *Nature, Lond.,* **262**: 102–107.

8.   *Mesochoerus limnetes* zone.   V. J. Maglio 1972.

9.   ER1470: F = 0·5%, 100F/P205 = 4·6, U = 8 ppm ($\equiv$eU$_3$O$_8$ = 13 ppm), N = nil.

11.   **ER1469.**   lt corpus mandibulae (f) with M3 (f) roots C–M2.
    ER1470.   cranium (f), roots dentes.
    ER1471.   proximal rt tibia (f).
    ER1472.   rt femur (f).
    ER1473.   proximal rt humerus (f).
    ER1474.   parietale (f).
    ER1475.   proximal and diaphysis rt femur (f).
    ER1481.   proximal and distal lt femur, tibia, distal lt fibula.
    ER1482.   corpus mandibulae (f), rt P4 (f), roots I2, C, P3 & M1, lt P3 (ff), P4 (f), M1 (ff), M2 (f), M3 (f), roots I2, C & M1.
    ER1483.   cranial fragment, lt corpus mandibulae (f); isolated M2.
    ER1801.   lt corpus mandibulae, C (root), P3 (root), P4, M1, M2 (roots), M3.
    ER1802.   corpus mandibulae, lt P4–M2, rt P3–M2.
    ER1803.   rt corpus mandibulae, roots of P3–M1.
    ER3230.   mandibula (crushed), all dentes.

12.   R. E. F. Leakey 1972, Men and sub-men on Lake Rudolf, *New Scient.,* **56** : 385–387 (ER1470).   R. E. F. Leakey 1973a, Further evidence of Lower Pleistocene hominids from East Rudolf, North Kenya, 1972, *Nature, Lond.,* **242** : 170–173 (remainder).   R. E. F. Leakey 1974, Further evidence of Lower Pleistocene hominids from East Rudolf, North Kenya, 1973, *Nature, Lond.,* **248** : 653–656 (ER1801, 1802, 1803).   R. E. F. Leakey 1976 (ER3230).   *Australopithecus.*

13.   R. E. F. Leakey and B. A. Wood 1974a, New evidence of the Genus *Homo* from East Rudolf, Kenya (IV), *Am. J. phys. Anthrop.,* **41** : 237–243 (ER1483).   R. E. F. Leakey and B. A. Wood 1974b, A hominid mandible from East Rudolf, Kenya, *Am. J. phys. Anthrop.,* **41** : 245–249 (ER1482).   M. H. Day, R. E. F. Leakey, A. C. Walker and B. A. Wood 1975, New hominids from East Rudolf, Kenya, I, *Am. J. phys. Anthrop.,* **42** : 461–476 (ER1470, ER1472, ER1481).   M. H. Day, R. E. F. Leakey, A. C. Walker and B. A. Wood 1976, New Hominids from East Turkana, Kenya, *Am. J. phys. Anthrop.,* **45**: 369–436 (ER1469, 1801, 1802, 1803, 1471, 1473).

15.   R. E. F. Leakey 1973b, Australopithecines and hominines: a summary of the evidence from the early Pleistocene of Eastern Africa, *in* S. Zuckerman (ed.) 1973, The Concepts of Human Evolution, *Symp. Zool. Soc. Lond.,* No. 33 : 53–69 (ER1470, 1481).   R. E. F. Leakey 1973a (ER1482).   R. E. F. Leakey 1974 (ER1802).   M. H. Day *et al.* 1975 (ER1470, ER1472, ER1481).   R. E. F. Leakey 1976 (ER3230).   M. H. Day *et al.* 1976 (ER1469, 1801, 1802, 1471).

16.   R. E. F. Leakey 1973b (ER1472, 1475, 1481).   R. E. F. Leakey 1976b, Hominids in Africa, *Am. Sci.,* **64** : 174–178 (ER1470).

KABUA

In September 1959, fragments of two individuals *(Homo sapiens)* were found by T. Whitworth, 1143 m SSE of Kabua water hole, Turkana District, in Upper Pleistocene lacustrine deposits, but they possibly represent more recent burials.   A preliminary report was published in 1960 in *Nature, Lond.,* **185** : 947–948.   A fuller account by T. Whitworth appeared in 1966, *S. Afr. archaeol. Bull.,* **21** : 138–150.

KANAM

1. Kanam.
2. In erosion gully at Kanam West, near the foot of Homa Mountain, about 750 m from the S shore of the Kavirondo Gulf, near Lake Victoria.   0° 21′ S, 34° 30′ E.
3. J. Gitau, assistant to L. S. B. Leakey, 29 March 1932, during East African Archaeological Research Expedition.
4. Possibly Kanam Beds.   Conference report 1933, Early Human Remains in East Africa, *Man,* **33** : 65–68.   L. S. B. Leakey 1935, *Stone Age Races of Kenya,* London : 15.   P. V. Tobias 1968, Middle and Early Upper Pleistocene Members of the Genus *Homo* in Africa *in* G. Kurth (ed.) 1968, *Evolution und Hominisation,* Stuttgart : 176–194.   Possibly calcrete block post-dating Kanam Beds.   K. P. Oakley 1975, A Reconsideration of the Date of the Kanam Jaw, *J. Archaeol. Sci.,* **2**: 151–152.
5. ————
6. Age uncertain.   P. G. H. Boswell 1935, Human Remains from Kanam and Kanjera, Kenya Colony, *Nature, Lond.,* **135** : 371.   P. V. Tobias 1962, A Re-examination of the Kanam Mandible, *in* G. Mortelmans and S. Nenquin (eds) 1962, *Proceedings of the 4th Pan-African Congress of Prehistory,* **1** : 341–360.   Lower Pleistocene.   L. S. B. Leakey 1935 : 18.   Middle Pleistocene.   P. V. Tobias 1968.   K. P. Oakley 1975.
7. ————
8. *Deinotherium bozasi, Stylohipparion albertensis, Ceratotherium simum, Hippopotamus imaguncula, Giraffa camelopardalis, Libytherium olduvaiensis.*   H. B. S. Cooke 1963, Pleistocene Mammal Faunas of Africa, with Particular Reference to Southern Africa, *in* F. C. Howell and F. Boulière (eds) 1963, African Ecology and Human Evolution, *Publs. Anthrop. Viking Fund,* vol. 36 : 78–84.
9. Mandible: F = 1·4–2·2%, 100F/P205 = 8–30, eU308 = 4–12 ppm.   Animal bones: F = 1·6–3·4%, 100F/P205 = 10–14, eU308 = 16–214 ppm.   K. P. Oakley 1975.
10. ————
11. **Kanam 1.**   adult: rt corpus mandibulae (ff), symphyseal region, rt P3–4.   Pathological.   **Holotype** of *Homo kanamensis* Leakey, 1935.   Plate 5.
12. L. S. B. Leakey 1932, The Oldoway Skeleton, *Nature, Lond.,* **129** : 721–722.
13. L. S. B. Leakey 1935.
14. P. V. Tobias 1962.   *Homo.*
15. P. V. Tobias 1962.
16. L. S. B. Leakey 1936, Fossil Human Remains from Kanam and Kanjera, Kenya Colony, *Nature, Lond.,* **138** : 643.   M. F. Ashley Montagu 1957, The chin of the

CATALOGUE OF
FOSSIL HOMINIDS

Key:

1. Place-name.
2. Location.
3. Discovery.
4. Geological deposit.
5. Evidence of burial.
6. Stratigraphical age.
7. Archaeological context.
8. Palaeontological context.
9. Analysis for relative age.
10. Absolute age.
11. Hominid remains.
12. First report.
13. First anatomical description.
14. Recent revision.
15. Best illustrations.
16. Other relevant publications.
17. Repository of fossils.
18. Repository of mould
　　　　　　　for casts.

: 335–338. P. V. Tobias 1960, The Kanam Jaw,
Stathopoulos 1975, Kanam Mandible's Tumour,

ory), Cromwell Road, London SW7 5BD,

Cromwell Road, London SW7 5BD, England.

KANAPOI

W and 14 km (9 miles) S of Teleki's volcano at the S
5° 04′ E.
m of Comparative Zoology, Harvard University, Ex-

minantly clastic but with layers of volcanic ash. B.
for Early Pleistocene Fossils in North-western Kenya,

adaurora, *Deinotherium bozasi*, *Parapapio jonesi*,
pattersoni, *N. plicatus*. B. Patterson, A. K.
70, Geology and Fauna of a New Pliocene Locality in
ond., **226** : 918–921. H. B. S. Cooke and R. F. Ewer
poi and Lothagam, North-western Kenya, *Bull. Mus.*
95.

BP (Geochron R-0554), 2·5 $\pm$ 0·2 Myr BP (Geochron
BP (Berkeley KA-2261) for the overlying basalt, which
to be in accord. However, it is not certain which
he faunal evidence is at variance with these dates.
4·5 Myr, based on correlation with Mursi (Yellow
sonable. B. Patterson *et al.* 1970.

13. . Howells 1967, Hominid Humeral Fragment from Early
estern Kenya, *Science, N.Y.*, **156** : 64–66.
14. H. , Early hominid humerus from East Rudolf, Kenya, *Science, N.Y.*
18
15. B. Pa. W. W. Howells 1967.
16. ———
17. National Museum, P.O. Box 40658, Nairobi.
18. National Museum, P.O. Box 40658, Nairobi.

KANJERA

1. Kanjera.
2. In erosion gullies in slopes about 550 m from the S shore of the Kavirondo Gulf, NE Lake Victoria.   0° 20′ S, 34° 30′ E.
3. L. S. B. Leakey and other members of the East African Archaeological Research Expediton.   19 March 1932 to 29 April 1932 (Kanjera 1–5), 1935 (Kanjera 3 part).
4. Surface and in grey calcified sands.   Kanjera 1–5.   Conference report 1933, Early Human Remains in East Africa, *Man*, **33** : 65–68.   L. S. B. Leakey 1935, *The Stone Age Races of Kenya*, London : 26–29.
5. ————
6. Middle Pleistocene.   L. S. B. Leakey 1935.   Geological age uncertain.   P. G. H. Boswell 1935, Human Remains from Kanam and Kanjera, Kenya Colony, *Nature, Lond.*, **135** : 371.
7. Chellean.   P. G. H. Boswell 1935.
8. *Simopithecus oswaldi, Elephas recki* (Stage 4), *Afrochoerus nicoli, Sivatherium oldowayensis, Pelorovis oldowayensis, Potamochoerus majus.*   L. S. B. Leakey 1972, *Homo sapiens* in the Middle Pleistocene and the evidence of *Homo sapiens'* evolution, *in* F. Bordes (ed.) 1972, *The Origin of Homo sapiens, Ecology and Conservation*, Unesco, No. 3 : 25–29.
9. Human cranial fragments *in situ:* eU308 = 15 ppm and 21 ppm; remainder from the surface: eU308 = 11–42 ppm.   Bones from Kanjera fauna: average of 5 samples eU308 = 135 ppm.   P. V. Tobias, 1968, Middle and Early Upper Pleistocene Members of the Genus *Homo* in Africa, *in* G. Kurth (ed.) 1968, *Evolution und Hominisation*, Stuttgart: 176–194.   Analyses indicate probability that these hominid remains are intrusive.   K. P. Oakley 1974, Revised Dating of the Kanjera Hominids, *J. hum. Evol.*, **3** : 257–258.
10. ————
11. **Kanjera 1.**   7 pieces of frontale, parietale, occipitale (ff).
    Kanjera 2.   3 pieces parietale (ff), costa (f).
    Kanjera 3.   8 pieces frontale, parietale, occipitale (ff), femur (ff), phalanx.
    Kanjera 4.   2 pieces frontale (ff).
    Kanjera 5.   2 femora (ff).
12. L. S. B. Leakey 1932, The Oldoway Skeleton, *Nature, Lond.*, **129** : 722.
13. L. S. B. Leakey 1935.   *Homo sapiens.*
14. ————
15. L. S. B. Leakey 1935.
16. P. G. H. Boswell 1935, Human Remains from Kanam and Kanjera, Kenya Colony, *Nature, Lond.*, **135** : 371.   L. S. B. Leakey 1936, Fossil Human Remains from Kanam and Kanjera, Kenya Colony, *Nature, Lond.*, **138** : 643.   W. W. Bishop 1963, The Later Tertiary and Pleistocene in Eastern Equatorial Africa, *in* F. C. Howell and F. Boulière (eds) 1963, African Ecology and Human Evolution, *Publs. Anthrop. Viking Fund*, No. 36 : 270.
17. British Museum (Natural History) Cromwell Road, London SW7 5BD, England.   Kanjera 1 and 3 (part) Reg. Nos M 16692–M 16709.   National Museum, P.O. Box 40658, Nairobi.   Kanjera 2, 3 (part), 4, 5.
18. British Museum (Natural History), Cromwell Road, London SW7 5BD, England.

KAPTHURIN

See under **BARINGO**

KOOBI FORA

See under **ILERET and KOOBI FORA**

LOBOI

1. Loboi.
2. Loboi localities II, VII and XV N of Lake Hannington (Bogoria), Baringo District. 0° 20′ N, 36° 05′ E.
3. University of Michigan excavations, 1974.
4. Loboi Silts, locality II (LB 160, LB 162, LB 163, LB 164, LB 165, LB 166), Locality VIII (LB 161), Locality XV (LB 159).   W. R. Farrand, in preparation.
5. Flexed burials with head pointing N or NW.
6. Late Pleistocene–Holocene.
7. Obsidian pieces in overlying Bogoria Silts.
8. Recent fauna.
9. ————
10. ————
11. **Loboi 1,** LB 159.   adult: rt corpus mandibulae (f), M1, M2.
    Loboi 2, LB 160.   adolescent, 15–17 yrs: lt ischium (f), pubis (f), lt femur (f), associated remains: 5 cranial fragments, lt corpus mandibulae (ff), vertebrae (ff), costae (ff), rt scapula (f), rt ilium (f), rt humerus (f), rt ulna (f), rt radius (f), rt caput femoris, lt fibula (f), rt talus, rt calcaneus, rt naviculare (f), 2 ossa cuneiforme, rt os cuboideum, ossa metatarsalia (f).
    Loboi 3, LB 161.   adult: skeleton (f).
    Loboi 4, LB 162.   adult male: skeleton (f) including rt parietale originally designated Loboi 5.
    Loboi 6, LB 163.   adult: lt temporale, 10 cranial fragments, processus coronoideus mandibulae.
    Loboi 7, LB 164.   adult: 9 cranial fragments.   27 post-cranial fragments.
    Loboi 8, LB 165.   adult: 16 cranial fragments, mandibular fragment, 2 shaft fragments.
    Loboi 9, LB 166.   adult: rt ischium (f).
12. M. H. Wolpoff, in preparation.
13. ————
14. ————
15. ————
16. ————
17. National Museum, P.O. Box 40658, Nairobi.
18. ————

LOTHAGAM

1. Lothagam.
2. On erosion slopes of Lothagam Hill (Locality 1), between the Kerio and Lomunyenkupurat Rivers just W of Lake Rudolf.   2° 53′ N, 36° 04′ E.
3. A. D. Lewis, assistant to B. Patterson, during Museum of Comparative Zoology, Harvard University, Expedition, 1967 (LT329); members of the Princeton University Expedition, 1972 (LT119 & 130).
4. Fluvio-deltaic deposits. B. Patterson, A. K. Behrensmeyer and W. D. Sill 1970, Geology and Fauna of a New Pliocene Locality in North-western Kenya, *Nature, Lond.,* **226** : 918.
5. ————
6. Pliocene.
7. ————
8. *Stegotetrabelodon orbus, Primelephas gomphotheroides, Tragelaphus* aff. *nakuae, Redunca* aff. *ancystrocera,* Pisces: *Lanistes carinatus, Etheria elliptica.*   B. Patterson *et al.* 1970.
9. ————
10. A3: *c.* 5·5 Myr BP on the basis of K/Ar dates of 8·31 ± 0·25 Myr BP (Berkeley JA-2294) for the underlying Pliocene volcanics and 3·71 ± 0·23 Myr BP (Berkeley KA-2262) for a basalt sill post-dating all the sedimentary deposits.   B. Patterson *et al.* 1970.
11. **LT119.**   occipitale (f).
    LT130.   rt parietale, rt temporale (f).
    LT329.   rt corpus mandibulae (f), M1 (i), roots M2–M3.
12. Anon. 1968, *New York Times* (LT329).
13. ————
14. ————
15. Anon. 1971, *Science News* : 99 (LT329).
16. ————
17. National Museum, P.O. Box 40658, Nairobi.
18. National Museum, P.O. Box 40658, Nairobi. Wenner-Gren Foundation for Anthropological Research, 14 East 71st Street, New York, N.Y. 10021, U.S.A. (mandible).

LUKEINO

1. Lukeino.
2. Erosion slope, Chepboit Locality 2/219, 5·5 km W of Yatya, Baringo District.   0° 50′ N, 35° 52′ E.
3. K. Chepboi working for M. Pickford, 1973.
4. Sand and grit-filled channel of Member A, Lukeino Formation, underlain by Kabarnet Trachyte Formation and overlain by the Kaparaina Basalt Formation.   M. Pickford 1975, Late Miocene sediments and fossils from the Northern Kenya Rift Valley, *Nature, Lond.,* **256** : 279–284.

5.   ——————
6.   Late Miocene.
7.   ——————
8.   Cercopithecoidea, Felidae, Canidae cf. *Crocuta*, *Enhydriodon* cf. *iluecai*, *Stegotetrabelodon orbus*, *Primelephas gomphotheroides*, *Deinotherium* cf. *bozasi*, *Hipparion turkanense*, *Nyanzachoerus tulotos*, *Hippopotamus*, *Giraffa* cf. *jumae*, *Orycteropus* spp. Reptilia, Aves, Pisces, Arthropoda, Mollusca. Fossil plants include leaves, wood, grass, ferns and diatoms. Algae are common. M. Pickford 1975.
9.   ——————
10. A3: *c.* 6·5 Myr on the basis of potassium-argon dating of the underlying Kabarnet Trachytes, 7·2 ± 0·3 Myr, 6·8 ± Myr and 6·7 ± 0·3 Myr and the overlying Kaparaina Basalt Formation, 5·4 ± 0·2 Myr. W. W. Bishop, G. R. Chapman, A. P. Hill and J. A. Miller 1971, Succession of Cainozoic Vertebrate Assemblages from Northern Kenya Rift Valley, *Nature, Lond.,* **233** : 389–394.
11. **KNM LU 335.** immature: lower lt M crown.
12. M. Pickford, 1975.
13. P. Andrews 1975, *in* M. Pickford 1975. Hominidae.
14. ——————
15. M. Pickford 1975.
16. ——————
17. National Museum, P.O. Box 40658, Nairobi.
18. National Museum, P.O. Box 40658, Nairobi.

LUKENYA HILL

1.   Lukenya Hill GvJm-22.
2.   Rock-shelter, base of hill, central southern face, Machakos district, 30 km SE of Nairobi. 1° 29′ S, 37° 03′ E.
3.   R. M. Gramly and workers, December 1970.
4.   1·40 m depth, 0·50 m below first pottery level, *in situ* with preceramic stone age industry, Horizon E. R. M. Gramly and G. P. Rightmire 1973, A fragmentary cranium and dated 'Later Stone Age' assemblage from Lukenya Hill, Kenya, *Man,* **8** : 571–579. R. Protsch 1975, The Absolute Dating of Upper Pleistocene Fossil Sub-Saharan Hominids and Their Place in Human Evolution, *J. hum. Evol.,* **4** : 297–322.
5.   ——————
6.   Upper Gamblian.
7.   Earliest 'Later Stone Age'. R. M. Gramly and G. P. Rightmire 1973.
8.   *Connochaetes taurinus*, *Alcelaphus buselaphus*, *Phacochoerus aethiopicus*, *Equus burchelli*, *Aepyceros melampus*, *Taurotragus oryx*. R. M. Gramly and G. P. Rightmire 1973.
9.   Hominid frontal: eU308 = 12 ppm, N = 0·08%. Fauna, *Connochaetes taurinus:* eU308 = 11·5 ppm, N = 0·14%. Fauna, *Taurotragus oryx:* eU308 = 12 ppm, N = 0·10%. R. Protsch 1975: 303.

10. A2: 17,700 $\pm$ 760 BP (UCLA-1709B), 17,650 $\pm$ 800 BP (UCLA-1709A) on basis of C 14 dating of collagen in bones of associated fauna.
    A1: *c.* 17,800 BP (5-UCLA-LJ) on basis of amino acid (aspartic acid) dating of hominid parietal.   R. Protsch 1975 : 315.   R. M. Gramly and G. P. Rightmire 1973.
11. **Lukenya Hill 1.**   adult male ?: frontale, lt parietale.
12. R. M. Gramly and G. P. Rightmire 1973.
13. R. Protsch 1975 : 316.   *H. sapiens afer.*
14. ————
15. ————
16. ————
17. National Museum, P.O. Box 40658, Nairobi.
18. National Museum, P.O. Box 40658, Nairobi.

MORUAROT HILL

Human material from Moruarot Hill, Turkana, of Later Stone Age (Wilton) was discovered by P. E. P. Deraniyagala in 1948 and is preserved in the National Museum, Colombo 7, Ceylon.   P. E. P. Deraniyagala 1965, *Spolia Zeylanica,* **30** : 188–203.

NAIVASHA

1. Naivasha.
2. Excavation below the Naivasha Railway Rock-shelter about 5 km from Naivasha on the Nairobi side and 100 m N of the main road.   0° 45′ S, 36° 26′ E.
3. A. J. Poppy, L. S. B. Leakey and M. D. Leakey, 11 July 1940.
4. Silts of the $\pm$ 73 m (200 ft) beach of Lake Naivasha.   L. S. B. Leakey 1942, The Naivasha Fossil Skull and Skeleton, *J. E. Africa nat. Hist. Soc.,* **16** : 169–177.
5. ————
6. Holocene.
7. Latest Upper Kenya Capsian D.   L. S. B. Leakey 1942.
8. Extant fauna.
9. ————
10. A1: 10,850 $\pm$ 300 BP (UCLA-1741) on basis of C 14 dating of collagen in hominid bone.   R. Protsch 1976, The Naivasha hominid and its confirmed Late Upper Pleistocene age, *Anthrop. Anz.,* **35** : 97–102.
11. **Naivasha 1.**   adult female over 50 years: cranium (f), mandibula (ff), ossa longa (f).
12. L. S. B. Leakey 1942.
13. L. S. B. Leakey 1942.   *Homo sapiens.*
14. ————
15. L. S. B. Leakey 1942.
16. ————
17. National Museum, P.O. Box 40658, Nairobi.   Reg. No. 22.
18. National Museum, P.O. Box 40658, Nairobi.

NGORORA

1. Ngorora.
2. On erosion slope about 0·5 km S of Bartabwa, Ngorora Location, Baringo District, NW of Lake Baringo.   0° 50′ N, 35° 50′ E.
3. G. R. Chapman, February 1968, whilst working for the East African Geological Research Unit, Bedford College, London.
4. Surface, said to be *ex* upper manganese palaeosol horizon (basal unit 'C') in the Ngorora Formation.   W. W. Bishop and G. R. Chapman 1970, Early Pliocene Sediments and Fossils from the Northern Kenya Rift Valley, *Nature, Lond.,* **226** : 914–918.   W. W. Bishop and M. H. L. Pickford 1975, Geology, fauna and palaeoenvironments of the Ngorora Formation, Kenya Rift Valley, *Nature, Lond.,* **254** : 185–192.
5. ————
6. Lower Pliocene.
7. ————
8. *Deinotherium hobleyi, Chilotheridium pattersoni, Brachypotherium, Protragocerus, ?Pseudotragus, Palaeotragus;* Reptilia: *Pelusios* cf. *sinuatus.*   W. W. Bishop, G. R. Chapman, A. Hill and J. A. Miller 1971, Succession of Cainozoic Vertebrate Assemblages from the Northern Kenya Rift Valley, *Nature, Lond.,* **233** : 389–394.   W. W. Bishop and M. H. L. Pickford 1975.
9. ————
10. A3: between 12 and 9 Myr on the basis of K/Ar dates for the fossiliferous horizons.   W. W. Bishop and G. R. Chapman 1970.
11. **BN1378.**   upper lt M2.
12. W. W. Bishop and G. R. Chapman 1970.
13. L. S. B. Leakey *in* W. W. Bishop and G. R. Chapman 1970.   Hominidae.
14. ————
15. W. W. Bishop and G. R. Chapman 1970.
16. ————
17. National Museum, P.O. Box 40658, Nairobi.
18. National Museum, P.O. Box 40658, Nairobi.

LIBYA

Charles Brian Montagu McBURNEY

Faculty of Archaeology and Anthropology,
University of Cambridge,
Cambridge,
England.

HAUA FTEAH

1. Haua Fteah.
2. Cave about 60 m above sea level, 8 km E of Apollonia, at foot of escarpment, about 500 m from shore.   32° 50′ N, 22° 10′ E.
3. Assistant to R. Hey, 1952; assistant to C. B. M. McBurney, 1955.
4. Margins of large hearth within occupation level in cave-earth.   Layer XXXIII.   C. B. M. McBurney 1961, Absolute Age of Pleistocene and Holocene deposits in the Haua Fteah, *Nature, Lond.,* **192** : 685–686.   C. B. M. McBurney 1967, *Haua Fteah and the Stone Age of the south east Mediterranean,* Cambridge: 1–387.
5. ————
6. Early Würm interstadial, cf. Brørup.   C. B. M. McBurney 1967.
7. Levalloiso-Mousterian.   C. B. M. McBurney, J. C. Trevor and L. H. Wells 1953a, The Haua Fteah Fossil Jaw, *Jl R. anthrop. Inst.,* **83** : 71–85.
8. *Ammotragus, Alcelaphus, Gazella* and other bovids, and *Equus.*   E. S. Higgs 1960, Some Mediterranean Faunas of the Mediterranean Coastal Areas, *Proc. prehist. Soc. (N.S.),* **27** : 144–154.   C. Emiliani 1964, Paleotemperature Analysis of Fossil Shells of Marine Molluscs, *in* H. Craig (ed.) 1964, *Isotopic and Cosmic Chemistry,* Amsterdam: 133–156.
9. Human mandibula 1: $eU308 = 8$ ppm, $N = 0\cdot2\%$.
    Human mandibula 2 : $F = 0\cdot07\%$, $100F/P205 = 0\cdot3$, $eU308 = 11$ ppm, $N = 0\cdot3\%$.   Animal bones from layers xxx–xxxiii : $F = 0\cdot03\%$, $100F/P205 = 0\cdot1$, $eU308 = 3$ ppm, $N = 0\cdot2\%$.
10. A2: 47,000 + 3200 − 2300 BP (GrN-2023) on basis of C14 dating of associated burnt bone.   J. C. Vogel & H. T. Waterbolk 1963, Groningen Radiocarbon Dates IV, *Radiocarbon,* **5** : 172.   C. B. M. McBurney 1961.
11. **Haua Fteah 1.**   female, 15–25 yrs: lt ramus and part lt corpus mandibulae (ff), M2, M3.
    Haua Fteah 2.   female, 15–20 yrs: lt ramus mandibulae (ff).
12. C. B. M. McBurney, J. C. Trevor and L. H. Wells 1953b, A fossil human mandible from a Levalloiso-Mousterian horizon in Cyrenaica, *Nature, Lond.,* **172** : 889–891.
13. C. B. M. McBurney *et al.* 1953a.
14. P. V. Tobias *in* C. B. M. McBurney 1967.
15. C. B. M. McBurney 1953a (Haua Fteah 1).   C. B. M. McBurney 1958, Evidence for the distribution in space and time of Neanderthaloids and allied strains in northern Africa, *in* G. H. R. von Koenigswald (ed.) 1958, *Hundert Jahre Neanderthaler,* Utrecht : 253–264, pl. 64.   (Haua Fteah 2).
16. C. B. M. McBurney 1960, *The Stone Age of Northern Africa,* London : 168, pl. 6.
17. University Museum of Anthropology and Archaeology, Downing Street, Cambridge, England.   Duckworth Laboratory Reg. Nos Af.8.7.1. and 2.
18. British Museum (Natural History), Cromwell Road, London SW7 5BD, England.

MALAWI

John Desmond CLARK

Department of Anthropology,
University of California,
Berkeley, California, 94720,
U.S.A.

FINGIRA

1. Fingira.
2. Rock-shelter on S side of Fingira Hill, SW end of Nyika Plateau.   About 10° 00′ S, 33° 25′ E.
3. B. H. Sandelowsky, 1966; K. R. Robinson, 1967.
4. Stratified cave deposit.   B. H. Sandelowsky and K. R. Robinson 1968, *Fingira: a Preliminary Report,* Department of Antiquities, Zomba, Malawi, Pub. No. 3.
5. Burials appear to have been made in the cave deposits over a period of time, the earlier having been disturbed and scattered by later interments.
6. Holocene.   B. H. Sandelowsky and K. R. Robinson 1968.
7. Late Stone Age.   Lithic assemblage comparable to those from Hora Mountain and Chenchere II. Affinities with Nachikufu Cave and Gwisho Springs in Zambia.
8. Extant fauna.   K. Olsen.   Unpublished report.
9. ―――――
10. A2: *c.* 3300 BP on basis of C14 dating of charcoals from 38–46 cm (15–18 inch) layer (3260 ± 80 BP UCLA-1258) and base of occupation deposit (3430 ± 80 BP UCLA-1259).   R. Berger and W. F. Libby 1968, (UCLA Radiocarbon Dates VII), *Radiocarbon,* **10** : 402–416.
11. **Fingira 1.**   male, 35–45 years: cranium, mandibula, vertebrae, costa (ff), 2 claviculae, 2 scapulae, 2 humeri, 2 radii, 2 ulnae, pelvis, 2 femora, 2 tibiae, 2 fibulae, pes.
    Fingira 2.   male, 35–45 years: partial cranium, mandibula, vertebrae (ff), clavicula (f) and scapulae (ff).
    Fingira unnumbered: fragmentary cranial and post-cranial bones (some complete) representing 5 infants (0–5 years), 3 children (6–19 years), 1 other immature individual, 2 adult males, 5 other adults representing a minimum number of 16 individuals.
12. B. H. Sandelowsky and K. R. Robinson 1968.
13. D. Brothwell and T. I. Molleson 1972, A study of the human skeletal remains of Late Stone Age date from Fingira Rock Shelter, Malawi, Unpublished report BM(NH).
14. ―――――
15. ―――――
16. J. D. Clark, C. V. Haynes, J. E. Mawby and A. Gautier 1970, Interim Report on Palaeo-Anthropological Investigations in the Lake Malawi Rift, *Quaternaria,* **13** : 305–354.   K. R. Robinson and B. H. Sandelowsky 1969, The Iron Age of northern Malawi: Recent work, *Azania,* **3** : 107–146.   B. M. Fagan and F. L. van Noten 1971, *The Hunter-gatherers of Gwisho,* Musée royal de L'Afrique Centrale, S. Humaines, Tervuren No. 74.
17. Sub-Department of Anthropology, British Museum (Natural History), London SW7 5BD.
18. ―――――

HORA

1. Hora.
2. Rock-shelter, Hora Mountain, about 48 km N of Mzimba.   11° 40′ S, 33° 45′ E.
3. J. D. Clark, 1950.
4. Stratified cave deposit.   L. H. Wells 1957, Late Stone Age Human Types in Central Africa, *Proc. 3rd Pan-Afr. Congr. Prehist.* : 183–185.
5. Two burials at 75 cm in 'Nachikufan II', stratum from level no higher than base of 'Nachikufan III'.   L. H. Wells 1953, *CHF* : 232.
6. Holocene.   L. H. Wells 1953.
7. Later Stone Age.   Nachikufan II/III.   L. H. Wells 1957.
8. Recent fauna.   L. H. Wells 1953.
9. —————
10. A3: probably 4000–1000 BP on basis of Nachikufan III assemblages from Nachikufu Cave and Leopard's Hill Cave in Zambia, dated between 1060 and 3500 BP, and from Fingira Rock-shelter, dated between 3260 and 3430 BP.
11. **Hora 1.** young male adult: calvaria, mandibula, vertebrae, costae (ff), clavicula, scapula (ff), 1 humerus, 2 radii, 2 ulnae, pelvis (ff), 2 femora, 2 tibiae, 2 fibulae (f), pes.
     Hora 2. young adult female: cranium, mandibula, vertebrae, costae (ff), 2 claviculae, 2 scapulae (f), 2 humeri, 2 radii, 2 ulnae, manus, pelvis (f), 2 femora, 2 tibiae, 2 fibulae (f), pes.
12. L. H. Wells 1953.   *Homo sapiens.*
13. J. D. Clark 1956, Prehistory in Nyasaland, *Nyasaland J., 9* : 92–119.
14. L. H. Wells 1957.
15. J. D. Clark 1959, *The Prehistory of Southern Africa,* London: Pl. 7.
16. C. Gabel 1963, Further Human Remains from the Central African Later Stone Age, *Man, 63* : 39.
17. Livingstone Museum, P.O. Box 498, Livingstone, Zambia.   Reg. No. 6988.
18. —————

MALI

Compiled by the Editors

ASSELAR

1. Asselar.
2. River terrace, 1·5 km W of Asselar military post, on margin of the Tilemsi depression, 20 km NE of In-Ourhi and about 400 km NE of Timbuktu.   18° 00′ N, 0° 15′ E.
3. M. V. Besnard and T. T. Monod, 20 December 1927.
4. Alluvial sands, lacustrine or flood deposit.
5. No positive evidence of burial.
6. Final Pleistocene or early Holocene.   T. Monod 1946, Sur l'âge de l'Homme d'Asselar, *Historia nat. Roma*, **1** : 81.   A. C. Blanc 1947, Sull'età geologica dell'Uomo di Asselar, *Riv. Antrop.*, **35** : 420.
    Neolithic.   L. Balout 1955, *Préhistoire de l'Afrique du Nord*, Paris : 81.
7. None associated, but microlithic sites nearby.   T. Monod 1946.
8. Pleistocene fauna indicative of pluvial conditions: *Lates niloticus, Crocodylus, Phacochoerus, Gazella* and freshwater and terrestrial mollusca including *Achatina*.   M. Boule and H. V. Vallois 1932, L'Homme Fossile d'Asselar (Sahara), *Archs Inst. Paléont. hum.*, **9** : 5–7.
9. Human costa: $F = 1·2\%$, $100F/P205 = 6·4$, $eU308 = 19$ ppm, $N = 0·5\%$.
10. A1: 6390 BP on basis of C14 dating of fragment of femur.   R. Mauny 1966, *in lit.*
11. **Asselar 1.**   adult male: skull and post-cranial skeleton.
12. Augiéras *et al.* 1931, D'Algérie au Sénégal, Mission Augiéras-Draper, 1927–8, *Soc. Géogr. Paris* : 255–257.
13. M. Boule and H. V. Vallois 1932.   Negroid.
14. T. Monod 1946.
15. M. Boule and H. V. Vallois 1932.
16. R. Mauny 1961, Catalogue des restes osseux humains préhistoriques trouvés dans l'Ouest africain, *Bull. Inst. fr. Afr. noire*, **23** (B) : 388–410.   T. Monod 1963, The late Tertiary and Pleistocene in the Sahara, *in* F. C. Howell and F. Bourlière (eds) 1963, African Ecology and Human Evolution, *Publs Anthrop. Viking Fund*, **36** : 147.
17. Institut de Paléontologie Humaine, 1 rue René Panhard, 75-Paris 13, France.
18. ———————

MAURITANIA

Compiled by the Editors

FORT GOURAUD   23° 43′ N, 12° 43′ W

In 1971 P. Elonard found human footprints being weathered out by the wind along the shores of an ancient (Tchadian) lake near Fort Gouraud.   Lacustrine clayey limestone inside a footprint has been dated by radiocarbon to 9120 ± 310 BP (Ly-489).

J. Evin, G. Marieu and C. Pachiaudi 1975, *Radiocarbon,* **17** : 23.

MOÇAMBIQUE

Based on data supplied by

Lereno BARRADAS

Instituto de Investigação Cientifica de Moçambique,
Lourenço Marques.

KASSIMATIS

1. Kassimatis.
2. Old quarry, Kassimatis, rt bank of Matola River, 400 m upstream from General Bettencourt's Bridge, Lourenço Marques.   25° 58′ S, 32° 26′ E.
3. L. Barradas, 3 October 1960.
4. Above 4 m above sea level in layer VI, estuarine or marine detrital pebbly and shelly sands or calcareous sandstone, with concretions.   L. Barradas 1965, *in lit.*
5. ————
6. Late Pleistocene.
7. Layer IV: Middle Stone Age.
8. Estuarine or marine molluscs including *Ostrea margaritacea.*   L. Barradas 1965, *in lit.*
9. ————
10. A3: *c.* 30,000 BP on basis of C14 dating of oyster shells from layer III: 33,720 $\pm$ 700 BP (SR-72).
11. **Kassimatis 1.**   adult: caput radii (ff).
12. L. Barradas, Concheiros de especial interesse no sul de Moçambique, *Mems Jta Inv est. Ultramar,* Lisb., in preparation.
13. ————
14. ————
15. ————
16. ————
17. Instituto de Investigação Cientifica de Moçambique, P.O. Box 1780, Lourenço Marques.
18. ————

MOROCCO

Pierre BIBERSON

Institut de Paléontologie Humaine,
1 rue René Panhard,
75-Paris 13,
France.

Emile ENNOUCHI

56 Avenue de la Résistance,
93-Le Raincy,
France.

&

Jean ROCHE

16 Avenue du Bel-Air,
75-Paris 12,
France.

'ALIYA

See under **MUGHARET EL-'ALIYA**

DAR ES-SOLTAN

1. Dar es-Soltan.
2. Cave, on the ' falaise morte ' of the Atlantic coast about 6 km SW of Rabat, 500 m S of an old summer-house of the Sultan.   34° 50′ N, 6° 20′ W.
3. A. Ruhlmann, 1937–1938.
4. Cave-earth (layer C1).   A. Ruhlmann 1951, La grotte préhistorique de Dar es-Soltan, *Hespéris,* **11** : 1–210.
5. Ibero-Maurusian burials into sterile layer.
6. Final Soltanian.   P. Biberson 1963, Human Evolution in Morocco in the Framework of the Paleoclimatic Variations of the Atlantic Pleistocene, *in* F. C. Howell and F. Bourlière (eds) 1963, African Ecology and Human Evolution, *Publs Anthrop. Viking Fund,* **36** : 439–442.
7. Ibero-Maurusian (= Oranian).   M. Antoine 1952, Les grandes lignes de la préhistoire marocaine, *Spec. Publ. 2nd Pan-Afr. Congr. Prehist.* : 240, Casablanca.
8. No fauna reported.
9. ———————
10. E. Ennouchi 1967, Gisements marocains datés par la Radioactivité, *Bull. Rech. Sci. Maroc.,* **11/12** : 118, Rabat.
11. **Dar es-Soltan 1.**   adult male: cranium (f), dentes, costae (ff), 3 metacarpalia, phalanges (ff), fibula, 1 cuneiforme, metatarsalia, phalanges.
    Dar es-Soltan 2.   young adult: calotte (ff), dentes.
12. H. V. Vallois 1951, Les restes humains de la grotte de Dar-es-Soltan, Appendix to Ruhlmann 1951.   Mechta-el-Arbi race.
13. D. Ferembach 1976, Les restes humains de la grotte de Dar-es-Soltan 2 (Maroc), Campagne, 1975, *Bull. Mém. Soc. Anthrop. Paris.,* Ser. 13, **3** : 183–193.
14. ———————
15. H. V. Vallois 1951.
16. C. B. M. McBurney 1960, *The Stone Age of Northern Africa,* London : 185–188.
17. Institut de Paléontologie Humaine, 1 rue René Panhard, 75 Paris 13, France.
18. ———————

JEBEL IRHOUD

1. Jebel Irhoud (formerly Ighoud).
2. Barytes mine, about 60 km SE of Safi.   31° 56′ N, 8° 52′ W.
3. Miner Mohammed Ben Fatmi, summer 1961: investigated by E. Ennouchi (Irhoud 1); C. S. Coon, 23 December 1963 (Irhoud 2).
4. Red sideritic clay, filling solution-cavity in Pre-Cambrian limestone.   E. Ennouchi 1962a, Un néandertalien: L'Homme du Jebel Irhoud (Maroc), *Anthropologie, Paris,* **66** : 279–299.

5. ———
6. Pre-Soltanian/Soltanian. P. Biberson 1964, La Place des Hommes du Paléolithique Marocain dans la Chronologie du Pléistocène Atlantique, *Anthropologie, Paris*, **68** : 475–526.
7. Levalloiso-Mousterian. E. Ennouchi 1963, Les Néanderthaliens du Jebel Irhoud (Maroc), *C.r. hebd. Séanc. Acad. Sci., Paris,* **256** : 2459–2460. Mousterian. L. Balout 1965, Le Moustérien du Maghreb, *Quaternaria,* **7** : 47.
8. Mammals of Middle Palaeolithic layers (i.e. Soltanian) include *Rhinoceros mercki, Equus mauritanicus, Gazella atlantica, Canis anthus, Bos primigenius.* E. Ennouchi 1962a. No mammals from skull site diagnostic of age. P. Biberson 1964.
9. ———
10. A2: > 30,000 yrs BP on basis of C14 dating of associated bone. E. Ennouchi 1966, Essai de datation du Gisement du Jebel Irhoud (Maroc), *Cr. Soc. Géol. France*: 405–406.
11. **Irhoud 1.** adult male: cranium (f) lacking basi-occipitale and dentes (base broken for brain extraction?).
     Irhoud 2. adult male: calvaria (ff) (frontale metopic).
     Irhoud 3. infant: mandibula (f), lt dc, C, dm2, M1; rt di2, roots of dm1 and dm2, M1.
12. E. Ennouchi 1962b, Un crâne d'Homme ancien au Jebel Irhoud (Maroc), *C.r. hebd. Séanc. Acad. Sci., Paris,* **254** : 4330–4332 (Irhoud 1). E. Ennouchi 1963 (Irhoud 2). E. Ennouchi 1969, Présence d'un enfant néanderthalien au Jebel Irhoud (Maroc), *Annls Paléont.,* (Vertébrés), **55** : 251–265 (Irhoud 3).
13. E. Ennouchi 1962a (Irhoud 1). Cf. *Homo neanderthalensis.* E. Ennouchi 1968a, Le Deuxième Crâne de l'Homme d'Irhoud, *Annls Paléont.,* (Vertébrés), **54** : 117–128 (Irhoud 2).
14. ———
15. E. Ennouchi 1962 (Irhoud 1). H. V. Vallois, in preparation (Irhoud 1 and 2).
16. C. Arambourg 1965, Le gisement moustérien et l'homme du Jebel Irhoud, *Quaternaria,* **7** : 1–7. E. Ennouchi 1968b, Empreintes des cerveaux des Néanderthaliens marocains. *Not. Mém. Serv. Géol. Maroc,* **29.** H. Suzuki 1970, *The Amud Man and his Cave Site,* The University of Tokyo: 191–2. R. Saban 1972, Les hommes fossiles de Maghreb, *Ouest médical,* **25** : 2443–2458. A. Mann and E. Trinkaus 1973, Neandertal and neandertal-like fossils from the upper Pleistocene, *Yrbk phys. Anthrop.,* **17** : 169–193. C. B. Stringer 1974, Population relationships of Later Pleistocene Hominids: A Multivariate Study of Available Crania, *J. archaeol. Sci.,* **1** : 317–342.
17. Laboratory of Geology, Université Mohammed V, Avenue Moulay-Cheriff, Rabat. (Temporarily at Institut de Paléontologie Humain, 1 rue René Panhard, 75-Paris 13, France.)
18. Muséum National d'Histoire Naturelle, 8 rue de Buffon, 75-Paris 5, France.

MUGHARET EL-'ALIYA

1. Mugharet el-'Aliya.
2. High Cave, Mugharet el-'Aliya, one of the Caves of Hercules, 8 km S of Cap Spartel, 13 km SW of Tangier. Approximately 35° 45′ N, 5° 56′ W.

3.   C. S. Coon, May 1939 ('Aliya 1 and 2); 1947 (two dentes).
4.   Not *in situ,* recovered from cave floor (Red 2, layer 9), presumed fallen from similar deposit in roof (Red 1, layer 5) ('Aliya 1).   *In situ* in red gritty cave earth (Red 2, layer 9) ('Aliya 2).   C. S. Coon 1940, *in* M. S. Senyürek 1940, Fossil Man in Tangier, *Pap. Peabody Mus.,* **16** 3: vii–viii.
5.   ————
6.   Ouljian/Soltanian.   P. Biberson 1961, Le Cadre Paléogéographique de la Préhistoire du Maroc Atlantique, *Publs. Serv. Antiq. Maroc.,* : 186–188.
7.   Layer 9: evolved Aterian; layer 5: final Aterian.   P. Biberson 1961.   L. Balout 1948, Les Fouilles américaines de la ' Grotte haute ' (Mougharet-el-Aliya, zone de Tanger) et la question S'Baikienne, *Bull. Soc. Hist. nat. Afr. N.,* **39** : 22–30.   B. Howe 1967, The Palaeolithic of Tangier, Morocco, *Am. Sch. prehist. Res. Peabody Mus. Harv. Univ. Bull.,* **22** : 1–200.
8.   North African Late Pleistocene fauna including layer 9: *Equus burchelli mauritanicus, Phacochoerus aethiopicus mauritanicus, Hippopotamus amphibius, Gazella dorcas;* Layer 5: *Equus burchelli mauritanicus, Hyaena spelaea, Giraffa camelopardalis, Gazella dorcas.*   B. Howe and H. L. Movius 1947, A Stone Age Cave Site in Tangier, *Pap. Peabody Mus.,* **28** : 21–32.
9.   ' Aliya 1 maxilla: $F = 0.03\%$, $100F/P205 = 0.1$, $N = 0.2\%$.   *Gazella dorcas* layer 5: $F = 0.03\%$, $100F/P205 = 0.1$, $N = 0.1\%$; layer 9: $F = 0.14\%$, $100F/P205 = 0.4$, $N = > 0.1\%$.   Conclusion: 'Aliya 1 derivation from layer 5 confirmed.   C. S. Coon 1962, *The Origin of Races,* London: 598.
10.   A3: *c.* 30,000 BP on basis of C14 dating of Upper Aterian charcoal from Dar es-Soltan: > 27,000 BP (UCLA–678B).
11.   **' Aliya 1.**   ' Tangier Man ', child about 9 yrs: part of lt maxilla and part lt malar; C, P, M2 (unerupted).
    ' Aliya 2.   ' Tangier Man ', adult: upper lt M2.
    1947: 2 dentes (not described).
12.   M. S. Senyürek 1940: 1–25.   *Homo neanderthalensis.*
13.   M. S. Senyürek 1940.
14.   L. C. Briggs 1955, The Stone Age Races of Northwest Africa, *Bull. Am. Sch. prehist. Res., Peabody Museum* (Cambridge, Mass.), **18** : 1–98.   C. S. Coon 1962: 598–600.
15.   M. S. Senyürek 1940.
16.   L. C. Briggs 1948, Les Hommes paléolithiques de Rabat et Tanger: Étude comparée, *Bull. Soc. Hist. nat. Afr. N.,* **39** : 105–114.   L. Balout 1955, Les Hommes Préhistoriques du Maghreb et du Sahara, *Publs Serv. Antiquités Algérie:* 199.   P. Biberson 1963, Human Evolution in Morocco in the Framework of the Paleoclimatic Variations of the Atlantic Pleistocene, *in* F. C. Howell and F. Bourlière (eds) 1963, African Ecology and Human Evolution, *Publs Anthrop. Viking Fund,* **36** : 441–442.
17.   Peabody Museum of Archaeology and Ethnology, Harvard University, Cambridge, Mass. 02138. U.S.A.   Reg. No. N/3635.
18.   ————

RABAT

1.   Rabat.
2.   Mifsud-Giudice quarry, Kébibat, on the coastal cliff SW of Rabat, near the Marie Feuillet hospital.   34° 02′ N, 6° 51′ W.

3. Workmen with M. Alenda, February 1933.   Subsequently examined by J. Marçais.
4. Consolidated dunes termed ' grès de Rabat ' with marine intercalations.   J. Marçais 1934, Découverte de Restes humains fossiles dans les grès quaternaires de Rabat (Maroc), *Anthropologie, Paris,* **44** : 579–583.   Upper part of great dune of Kébibat.   P. Biberson 1963, Human Evolution in Morocco in the Framework of the Paleo-climatic Variations of the Atlantic Pleistocene, *in* F. C. Howell and F. Bourlière (eds) 1963, African Ecology and Human Evolution, *Publs Anthrop. Viking Fund,* **36** : 435.
5. ─────────
6. Post-Tyrrhenian/Pre-Grimaldian (Riss-Würm).   R. Neuville and A. Ruhlmann 1942, L'Age de l'Homme fossile de Rabat, *Bull. Mém. Soc. Anthrop. Paris,* **3,** 9 : 74–88.   Pre-Tyrrhenian.   G. Choubert and J. Marçais 1947, Le Quaternaire des environs de Rabat et l'âge de l'Homme de Rabat, *C.r. hebd. Séanc. Acad. Sci., Paris,* **224** : 1645–1647.   Dating theories discussed *in* G. Lecointre 1960, Le gisement de l'Homme de Rabat, *Bull. Archéol. Maroc,* **3** : 55–85.   Middle Tensiftian.   P. Biberson 1963.
7. No associated industry.   Probably contemporaneous with Acheulian Stage VII.   P. Biberson 1964, La Place des Hommes du Paléolithique Marocain dans la Chronologie du Pleistocene Atlantique, *Anthropologie, Paris,* **68** : 475–526.
8. *Elephas atlanticus, Rhinoceros simus, Equus mauritanicus, Bos, Alcelaphus, Gazella.*   C. Arambourg 1938, Mammifères fossiles du Maroc, *Mém. Soc. Sci. nat. phys. Maroc,* **46** : 1–74.   E. Ennouchi 1948, Les Mammifères quatérnaires de Rabat, *Soc. Sci. Nat. Rabat,* **2.**   E. Ennouchi 1952, Sur un ensemble de nouvelles pièces paléontologiques de la faune de Rabat., *C.r. Soc. Sci. Nat. Maroc*: 129–131, Rabat.
9. Rabat 1 : F = 1·4%, 100F/P205 = 6·1.
    Mammal bones: F = 1·8%, 100F/P205 = 7·3; F = 1·4%, 100F/P205 = 4·5.
10. A3: > 200,000 BP on basis of Th230/U234 in shells of sample L–841T from over-lying layer.   C. E. Stearns and D. L. Thurber 1965, Th230-U234 dates of late Pleistocene marine fossils from the Mediterranean and Moroccan littorals, *Quaternaria,* **7** : 29–42.
11. **Rabat 1.**   ' Rabat Man ', adolescent male: fragments of calotte, lt maxilla (ff) with 2I, C, 2P, M1–2; mandibula (f), lt 2I, C, P3–M1, rt I1, P3–M3.
12. J. Marçais 1934.
13. H. V. Vallois 1945, L'Homme fossile de Rabat, *C.r. hebd. Séanc. Acad. Sci., Paris,* **221** : 669–671.
14. R. Saban 1975, Les restes humains de Rabat (Kebibat), *Annls. Paléont.,* (Vertébrés), **61** : 153–207.
15. H. V. Vallois 1960.
16. C. Arambourg and P. Biberson 1956, The Fossil Human Remains from the Paleolithic site of Sidi Abderrahman (Morocco), *Am. J. phys. Anthrop.,* **14** : 467–489.   L. C. Briggs 1948, Les Hommes paléolithiques de Rabat et Tanger: étude comparative, *Bull. Soc. Hist. nat. Afr. N.,* **39** : 105–114.   H. V. Vallois 1960, L'Homme de Rabat, *Bull. Archéol. maroc.,* **3** : 87–91.   Pre-neandertalian of Africa.   R. Saban 1972, Les hommes fossiles du Maghreb, *Ouest médical,* **25** : 2443–2458.
17. Institut de Paléontologie Humaine, 1 rue René Panhard, 75 Paris 13, France.
18. ─────────

SALÉ

1. Salé.
2. El Hamra, near Douar Caïd bel Aroussi.   34° 04′ N, 6° 46′ W.
3. Quarry-men.   Collected by J. J. Jaeger, July 1971.
4. Aeolian sands of Salé dune.   J. J. Jaeger 1975, The mammalian faunas and hominid fossils of the Middle Pleistocene of the Maghreb, *in* K. W. Butzer and G. Ll. Isaac (eds) 1975, *After the Australopithecines* : 399–418, The Hague.
5. ————
6. Anfatian.   J. J. Jaeger 1975.
7. No archaeological remains.
8. *Ceratotherium simum, Equus* cf. *mauritanicus, Connochaetes,* canid indet.   Mollusca.
9. ————
10. ————
11. **Salé.**   adult: calvaria, lt maxilla with I2–M2, rt M3, natural endocranial cast (f).
12. J. J. Jaeger in press, Découverte d'un crâne d'hominidé dans le Pleistocène moyen du Maroc, *in* Evolution des Vertébrés, *Colloque International du CNRS,* Paris.
13. J. J. Jaeger 1973, Un Pithécanthrope évolué, *Recherche,* **4** : 1006–1007.
14. ————
15. J. J. Jaeger 1973.
16. ————
17. Collections du Service Géologique du Maroc, Rabat.
18. Laboratoire de Paléontologie, Université des Sciences et Techniques du Languedoc, 34060–Montpellier, France.

SIDI ABDERRAHMAN

1. Sidi Abderrahman.
2. ' Grotte des Littorines ', exposed in Schneider gravel pit, 6 km SW of Casablanca.   33° 35′ N, 7° 40′ W.
3. P. Biberson, 20 March 1955.
4. Karstic cavity in consolidated fine-grained calcareous sandstone lens F, in continuity with level Do at cave entrance.   C. Arambourg and P. Biberson 1956, The fossil human remains from the Paleolithic site of Sidi Abderrahman (Morocco), *Am. J. phys. Anthrop.,* **14** : 467–489; R. Neuville and A. Ruhlmann 1941, La place du Paléolithique ancien dans le quaternaire marocain, *Hespéris,* **8** : 1–156.
5. ————
6. Early Tensiftian.   P. Biberson 1956, Le Gisement de l'Atlanthrope de Sidi Abderrahman (Casablanca), *Bull. Archéol. maroc.,* **1** : 39–92.   P. Biberson 1961a, Le Cadre Paléogéographique de la Préhistoire du Maroc atlantique, *Publs. Serv Antiq. Maroc,* **16** : 159–164.
7. Upper Middle Acheulian, Stage VI.   P. Biberson 1956 : 65–80.   P. Biberson 1961b, Le Paléolithique inférieur du Maroc Atlantique, *Publs Serv. Antiq. Maroc,* **17** : 283–286.
8. *Crocuta crocuta, Bos primigenius, Alcelaphus bubalis, Gazella, Rhinoceros simus, Equus mauritanicus.*   P. Biberson 1956 : 61.

9.   ————
10. ————
11. **Sidi Abderrahman 1.** ' Casablanca Man '. young adult male: part of rt mandibula
       (f), M1, M2, M3; lt mandibula (f), P3.
12. C. Arambourg and P. Biberson 1955, Découverte de vestiges humains acheuléens dans
       la carrière de Sidi Abderrahman près Casablanca, *C.r. hebd. Séanc. Acad. Sci., Paris,*
       240 : 1661–1663.
13. C. Arambourg and P. Biberson 1956. Cf. *Atlanthropus.*
14. ————
15. C. Arambourg and P. Biberson 1956.
16. C. Arambourg 1957, Récentes découvertes de paléontologie humaine réalisées en
       Afrique du Nord française *(L'Atlanthropus* de Ternifine—L'Hominien de Casablanca),
       Proc. 3rd Pan-Afr. Congr. Prehist. : 186–194.   P. Biberson 1963, Human Evolution in
       Morocco in the Framework of the Paleoclimatic Variations of the Atlantic Pleistocene,
       in F. C. Howell and F. Bourlière (eds) 1963, African Ecology and Human Evolution,
       Publs Anthrop. Viking Fund, **36** : 417–447.   P. Biberson 1964, La place des Hommes
       du Paléolithique marocain dans la Chronologie du Pleistocene atlantique,
       Anthropologie, Paris **68** : 475–526.   R. Saban 1972, Les hommes fossiles du Maghreb,
       Ouest médical, **25** : 2443–2458.
17. Muséum Nationale d'Histoire Naturelle, 8 rue de Buffon., 75-Paris 5, France.
18. Muséum Nationale d'Histoire Naturelle, 8 rue de Buffon, 75-Paris 5, France.

TAFORALT

1.   Taforalt (1951).
2.   Cave, ' Grotte des Pigeons ' at foot of cliff overlooking Zegrel valley, near the Oujda-
       Moulouia route; 500 m E of Taforalt village, 55 km NW Oujda, in the Beni-Snassen
       mountains.   34° 45′ N, 2° 15′ W.
3.   J. Roche, 1951.
4.   Ashy occupation layer in yellow earths (Niveau D).   J. Roche 1953, La Grotte de Ta-
       foralt, *Anthropologie, Paris,* **57** : 374–380.
5.   ————
6.   Late Soltanian.   P. Biberson 1961, Le Cadre Paléogéographique de la Préhistoire du
       Maroc atlantique, *Publs Serv. Antiq. Maroc.,* **16** : 189.
7.   Late Aterian.   J. Roche 1953.
8.   Sparse fauna includes *Phoca.*   J. Roche 1953.
9.   ————
10. A3: *c.* 21,000 BP on basis of C14 dating of Late Aterian site at Fachi, Niger: 21,350 ±
       350 BP (T–340B).
11. **Taforalt 1.**   parietale (ff).
12. J. Roche 1953.
13. D. M. Ferembach 1962, La Nécropole épipaléolithique de Taforalt, Étude des
       Squelettes humains, *Publ. Centre nat. Rech. Sci.,* Rabat, Appendix : 123.
14. ————
15. ————

16. L. Balout 1955, Les Hommes Préhistoriques du Maghreb et du Sahara, *Serv. Antiq. Algerie,* Algiers : 32.   D. Ferembach 1976a, Les restes humains atériens de Témara (Campagne 1975), *Bull. Mém. Soc. Anthrop. Paris,* Ser. 13, **3** : 175–180.   D. Ferembach 1976b, Les restes humains de la grotte de Dar-es-Soltan 2 (Maroc), Campagne 1975, *Bull. Mém. Soc. Anthrop., Paris.,* Ser. 13, **3** : 183–193.
17. Institut de Paléontologie Humaine, 1 rue René Panhard, 75-Paris 13, France.
18. ─────────

1. Taforalt (1951–1953).
2. Cave, ' Grotte des Pigeons ' at foot of cliff overlooking Zegzel valley, near the Oujda-Moulouia route; 500 m E of Taforalt village, 55 km NW of Oujda in the Beni-Snassen mountains.   34° 45′ N, 2° 15′ W.
3. J. Roche, 1951 (Niveau A – Taforalt 3 and 4); 1952–53 (Niveau B – Taforalt 5 onwards); R. P. Bienvenu-Blondeau, 1952 (Niveau B – Taforalt 2).
4. Ashy occupation layers in stony beds (Niveau A–C).   J. Roche 1953, La Grotte de Taforalt, *Anthropologie, Paris,* **57** : 375–380.
5. Burials into Niveau B.   J. Roche 1953.
6. Final Soltanian; deposits equivalent to Flandrian stage.   P. Biberson 1961, Le Cadre Paléographique de la Préhistoire du Maroc atlantique, *Publs Serv. Antiq. Maroc,* **16** : 189.
7. Ibero-Maurusian (= Oranian).   J. Roche 1963, *L'Épipaléolithique marocain,* Lisbon.
8. Recent fauna abundant in Oranian layers.   J. Roche 1958–59, L'Épipaléolithique marocain, *Libyca,* **6–7** : 159–192.
9. ─────────
10. A2: on Niveau II: 10,800 $\pm$ 400 BP (Sa–13); VI: 12,070 $\pm$ 400 BP (Sa–14); VIII: 10,500 $\pm$ 400 BP (Sa–15); 11,900 $\pm$ 240 BP (L–399E).   J. Roche 1959, Nouvelle datation de l'épipaléolithique marocain par la methode du carbone 14, *C.r. hebd. Séanc. Acad. Sci., Paris,* **249** : 729–730.
11. **Taforalt 2.**   skeleton (ff).
     Taforalt 3.   cranium (ff).
     Taforalt 4.   fragments of skeleton.
     Taforalt 5.   ossuary burials: remains of 180 adults including 39 males, 31 females; 10 sex not determined; 6 adolescents; 97–100 children.
12. J. Roche 1955, Note préliminaire sur la Grotte de Taforalt (Maroc oriental), *Proc. 2nd Pan-Afr. Congr. Prehist.,* : 647–652.
13. D. M. Ferembach 1962, La Nécropole épipaléolithique de Taforalt, Étude des Squelettes humains, *Publ. Centre nat. Rech. Sci.,* Rabat.   Appendix : 123.
14. M. C. Chamla 1970, Les Hommes Épipaléolithiques de Columnata (Algérie Occidentale), *Mém. Cent. Rech. Anthrop. Préhist. Ethnogr. Alger,* **15** : 1–132.
15. D. M. Ferembach 1962.
16. D. M. Ferembach 1960, Les Hommes du Mesolithique d'Afrique du Nord et le Problème des isolats, *Bolm Soc. port. Cienc. nat.,* **23** (2) : 1–16.   D. M. Ferembach 1959, Les Restes humains épipaléolithiques de la Grotte de Taforalt (Maroc oriental), *C.r. hebd. Séanc. Acad. Sci., Paris,* **248** : 3465–3467.   P. Biberson 1963, Human

Evolution in Morocco in the Framework of the Paleoclimatic Variations of the Atlantic Pleistocene, *in* F. C. Howell and F. Bourlière (eds) 1963, African Ecology and Human Evolution, *Publs Anthrop. Viking Fund,* **36** : 441–442.   R. Saban 1972, Les hommes fossiles du Maghreb, *Ouest médical,* **25** : 2443–2458.
17. Institut de Paléontologie Humaine, 1 rue René Panhard, 75 Paris 13, France.
18. ————————

' TANGIER MAN '

See under **MUGHARET EL-'ALIYA**

TÉMARA

1. Témara.
2. Cave, ' Grotte des Contrebandiers ', in a coastal cliff, 17 km SW of Rabat, 53 km NE of Casablanca.   Approximately 33° 55′ N, 7° 00′ W.
3. J. Roche, 1956.
4. Isolated block of grey-pink breccia.   P. Biberson 1961, Le Cadre Paléogéographique de la Préhistoire du Maroc atlantique, *Publs Serv. Antiq. maroc,* **16** : 176.
5. ————————
6. Epi-Ouljian.   P. Biberson 1961.
7. Upper Acheulian.   H. V. Vallois and J. Roche 1958, La Mandibule acheuléenne de Témara, Maroc, *C.r. hebd. Séanc. Acad. Sci., Paris,* **246** : 3113–3116.
   Acheulian stage VIII.   P. Biberson 1961.
8. None reported.
9. ————————
10. ————————
11. **Témara 1.**   ' Témara Man ': corpus mandibulae (f), dentes and dentes maxillares.
12. H. V. Vallois and J. Roche 1958.   Pre-neandertalian.
13. ————————
14. ————————
15. ————————
16. H. V. Vallois 1960, L'Homme de Rabat, *Bull. Archéol. maroc,* **3** : 87–91.   P. V. Tobias 1968, Middle and early Upper Pleistocene Members of the genus *Homo* in Africa, *in* G. Kurth (ed.) 1968, *Evolution und Hominisation,* Stuttgart: 184.   J. Roche 1963, L'Épipaléolithique marocain, Lisbon : 190–192.   P. Biberson 1964, La Place des Hommes du Paléolithique Marocain dans le Chronologie du Pléistocène atlantique, *Anthropologie, Paris,* **68** : 475–526.   R. Saban 1972, Les hommes fossiles du Maghreb, *Ouest médical,* **25** : 2443–2458.   J. Roche 1976, Chronostratigraphie des restes atériens de la grotte des contrebandiers à Témara (Province de Rabat), *Bull. Mém. Soc. Anthrop., Paris,* Ser. 13, **3** : 165–173.
17. Institut de Paléontologie Humaine, 1 rue René Panhard, 75-Paris 13, France.
18. ————————

THOMAS QUARRIES

1. Thomas Quarries.
2. Quarry I (Thomas 1) and Quarry III (Thomas 2) S of Casablanca, 8·5 km from the Casablanca to Moulay-bou-Chaib and Azemmour coastal route.   33° 30′ N, 7° 45′ W.
3. P. Beriro, 13 April 1969 (Thomas 1); E. Ennouchi and P. Beriro, 1972 (Thomas 2).
4. Calcareous aeolian sandstones, 9·50 m below the upper level of the cutting front and 2·50–3·0 m above the middle level.   E. Ennouchi 1969, Découverte d'un Pithécanthropien au Maroc, *C.r. hebd. Séanc. Acad. Sci., Paris,* **269 D** : 763–765 (Thomas 1).   Calcite concretions in small cave (Thomas 2).   E. Ennouchi 1972, Nouvelle découverte d'un Archanthopien au Maroc, *C.r. hebd. Séanc. Acad. Sci., Paris,* **274 D** : 3088–3090.
5. ————
6. Amirian.   Corresponds to dune H at Sidi-Abderrahman.   E. Ennouchi 1969.   E. Ennouchi 1972.
7. ————
8. Quarry I: *Elephas iolensis, Elephas* sp., *Equus mauritanicus, Canis anthus, Hyaena crocuta spelaea, Ursus arctos bibersoni, Bos primigenius, Alcelaphus bubalis, Gazella atlantica, Hystrix cristata, Lepus kabylicus, Hippopotamus amphibius, Phacochoerus africanus;* Reptilia: *Testudo;* Aves: *Struthio.* Quarry III: *Rhinocerus simus* in addition. E. Ennouchi 1970, Un nouvel archanthropien au Maroc, *Annls. Paléont. (Vertébrés),* **56** : 95–107.
9. ————
10. ————
11. **Thomas 1.**   adult: lt ramus mandibulae (f), P4–M3.
    Thomas 2.   female or young adult: rt frontale, temporale, orbita, maxilla.
            upper lt; I2, 2C, 2P3, 2P4, 2M1, M2, M3 (erupting).
12. E. Ennouchi 1969 (Thomas 1).   *Atlanthropus mauritanicus.*   E. Ennouchi 1972 (Thomas 2).   *Atlanthropus mauritanicus.*
13. E. Ennouchi 1970 (Thomas 1).   E. Ennouchi 1973, Un nouvel archanthropien à la carrière Thomas (Maroc), *Annls Paléont. (Vertébrés),* **56** : 95–105.   (Thomas 2).
14. F. Sausse 1975, La mandibule atlanthropienne de la Carrière Thomas I (Casablanca), *Anthropologie, Paris,* **79** : 81–112.   E. Ennouchi 1976, Le deuxième Archanthropien à la carrière Thomas 3 (Maroc), *Bull. mus. Nat. Paris,* in press.
15. E. Ennouchi 1970 (Thomas 1).   E. Ennouchi 1973 (Thomas 2).   F. Sausse 1975.
16. R. Saban 1972, Les hommes fossiles du Maghreb, *Ouest médical,* **25** : 2443–2458.
17. At present Musée de l'Homme, Palais de Chaillot, 75-Paris 16, France.
18. ————

NIGER

Compiled by the Editors

ACHEGOUR   18° 51′ N, 11° 36′ E.

Between July and September 1960, G. Popov discovered surface remains of at least 2 skeletons at an open site near Salvador, NW of Bilma, 50 km W, on the car track from l'Arbre du Ténéré.   The remains are believed to be of Holocene age.   Blade tools of silcrete lay on the surface nearby (cf. Upper Palaeolithic of Europe).

R. Mauny 1961, Catalogue des Restes osseux humains préhistoriques trouvés dans L'Ouest africain, *Bull. Inst. fr. Afr. noire*, **23**(B) : 388–410.

The remains are preserved at the British Museum (Natural History), Cromwell Road, London SW7 5BD, England.   Reg. Nos EM604–5.

NIGERIA

Charles Thurstan SHAW

Silver Ley,
37 Hawthorne Road,
Stapleford,
Cambridgeshire,
England.

IWO ELERU

1. Iwo Eleru.
2. Rock shelter 24 km (15 miles) from Akure, Western Nigeria.   7° 27′ N, 5° 08′ E.
3. T. Shaw, 1965.
4. Rock surface in cavity formed by boulders.
5. Contracted burial, probably Late Stone Age.
6. Late Pleistocene.
7. Late Stone Age. T. Shaw 1965, Akure excavations, *African Notes,* Ibadan, **3** : 5–6. T. Shaw 1966, The Late Stone Age in the Nigerian Forest, *Etudes et documents Tchadiens,* Mem. **1** : 364–373.
8. None reported.
9. Human cranium: eU308 = 25 ppm, N = nil.
10. A2: 11,200 ± 200 BP (I–1753) on the basis of C14 dating of closely associated charcoal: *Radiocarbon,* **10** : 289–290.
11. **Iwo Eleru 1.** adult, over 35 years, male: calotte, mandibula (f) with 30 isolated dentes, including 4C, 4P, 6M, all much worn, ossa longa (f) without epiphyses, and other small fragments.
12. T. Shaw 1965.
13. D. R. Brothwell and T. Shaw 1971, A Late Upper Pleistocene Proto-West African negro from Nigeria, *Man,* **6** : 221–227.
14. —————
15. D. R. Brothwell and T. Shaw 1971 : Pls 6b, 7a, 7b.
16. —————
17. University of Ibadan, Ibadan.
18. Sub-Department of Anthropology, British Museum (Natural History), Cromwell Road, London SW7 5BD, England.

RHODESIA

David Walter PHILLIPSON

British Institute in Eastern Africa,
P.O. Box 30710,
Nairobi,
Kenya.

INYANGA

1. Inyanga.
2. Rockshelter at Chitura Rocks (site XXVII), 6 km E of the confluence of Nyabombgwe and Nyagombe Rivers, Inyanga District.   17° 44′ S, 32° 18′ E.
3. Excavations by K. R. Robinson, 1951.
4. Stratified cave deposit.
5. In shallow grave sealed by ochreous earth layer containing exclusively Wilton artifacts.
6. Holocene.   K. R. Robinson 1958, Some Stone Age Sites in Inyanga District, *in* R. Summers 1958, *Inyanga,* Cambridge : 270–309.
7. Late Stone Age, cf. Wilton.   K. R. Robinson 1958.
8. Extant fauna with extinct *Equus* cf. *harrisi.*   H. B. S. Cooke 1958, Fossil Animal Remains from Inyanga, *in* R. Summers 1958 : 152–158.
9. ─────────
10. ─────────
11. **Inyanga 1.**   cranium, mandibula, all dentes except rt Il, atlas, axis (f).   Rest of skeleton was present in the grave but was not preserved for study.
12. K. R. Robinson 1958.
13. P. V. Tobias 1958, Skeletal Remains from Inyanga, *in* R. Summers 1958 : 159–172.   'Bushman'.
14. ─────────
15. P. V. Tobias 1958.
16. ─────────
17. National Museum of Rhodesia, P.O. Box 240, Bulawayo.   Reg. No. 7418.
18. ─────────

SOUTH AFRICA

Phillip Vallentine TOBIAS

&

Kay COPLEY

University of the Witwatersrand, Johannesburg,
Medical School,
Hospital Street,
Johannesburg,
Transvaal.

&

Charles Kimberlin BRAIN

Transvaal Museum,
Pretoria,
Transvaal.

BORDER CAVE

1. Border Cave.
2. Cave about 90 m below the crest of the Lebombo range scarp, 365 m on the KwaZulu (Zululand) side of the international boundary with Swaziland, 13 km N of Ingwavuma Village, Natal.   Approximately 27° 01′ S, 31° 59′ E.
3. D. Drew, January 1941 (Border Cave 1–partim).   T. R. Jones, 1941 Border Cave 1—partim–parietal); H. B. S. Cooke, B. D. Malan and L. H. Wells, July 1941 and July 1942 (Border Cave 1–partim–Border Cave 2 and 3); C. Powell and P. B. Beaumont, April 1974 (Border Cave 4).
4. Border Cave 1 and 2 found in dump resulting from W. E. Horton's pit.   H. B. S. Cooke, B. D. Malan and L. H. Wells 1945, Fossil Man in the Lebombo Mountains, South Africa: The 'Border Cave', Ingwavuma District, Zululand, *Man,* **45** : 6–13.   Third White Ash (Border Cave 4).
5. Border Cave 3: burial in shallow grave, located in dark earth layer of ' normal Pietersburg ' zone; inferred to have been buried from a layer not higher than an ash horizon (2nd White Ash ?) at the base of the overlying advanced or ' Epi-Pietersburg ' stratum.   H. B. S. Cooke, B. D. Malan and L. H. Wells 1945.
6. Upper Pleistocene.   H. B. S. Cooke, B. D. Malan and L. H. Wells 1945.
7. Border Cave 1, 2 and 3: inferred to belong to Epi-Pietersburg (= Final Middle Stone Age) or, possibly, even to normal Pietersburg (= Full Middle Stone Age).   H. B. S. Cooke, B. D. Malan and L. H. Wells 1945.   L. H. Wells 1950, The Border Cave Skull, Ingwavuma District, Zululand, *Am. J. phys. Anthrop.,* **8** : 241–243.   L. H. Wells 1959, The problem of Middle Stone Age man in Southern Africa, *Man,* **59** : 158–160.   P. B. Beaumont and J. C. Vogel 1972, On a new radiocarbon chronology for Africa south of the equator, *African Studies,* **31** : 65–89 and 155–182.   P. B. Beaumont 1973, Border Cave – a progress report, *S. Afr. J. Sci.,* **69** : 41–46.   Middle Stone Age (Border Cave 4).   H. de Villiers 1976, A second adult human mandible from Border Cave, Ingwaruma District, KwaZulu, South Africa, *S. Afr. J. Sci.,* **72** : 212–215.
8. Modern fauna of lowveld facies except *Equus kuhni* and an extinct bovine.   H. B. S. Cooke, B. D. Malan and L. H. Wells 1945.
9. Border Cave 1: N = 0·28% (BMNH): 0·44% (UCLA).
    Border Cave 2: N = 0·29% (BMNH).
    Border Cave 3: N = 0·44% (BMNH); 0·41% (UCLA).
    Animal bones from Middle Stone Age levels: N = 0·24–0·85% (BMNH).
10. A1: *c.* 60,000 BP (2-UCLA-LJ) on basis of racemization of isoleucine in Border Cave 3. R. Protsch 1975, The Absolute Dating of Upper Pleistocene Subsaharan Fossil Hominids and Their Place in Human Evolution, *J. hum. Evol.,* **4** : 297–322. A2: > 48,700 BP (Pta–489) and > 48,350 BP (Pta–459) on basis of C14 dating of charcoals from the Epi-Pietersburg stratum.   J. C. Vogel and P. B. Beaumont 1972, Revised radiocarbon chronology for the Stone Age in South Africa, *Nature, Lond.,* **237** : 50–51.   P. B. Beaumont and J. C. Vogel 1972.   Possibly 80,000–100,000 BP. H. de Villiers 1976.
11. **Border Cave 1.** adult male?: frontale, parietalia (ff), temporalia, occipitale (ff), rt zygomaticum (f).
    Border Cave 2. adult female ?: mandibula without dentes (f).
    Border Cave 3. infant about 4 to 6 months: cranium (ff), mandibula, dentes (f), longbone diaphyses, costae, vertebrae, ilia, ischium, scapulae, clavicula, os metacarpalia, talus and calcaneus.

Border Cave 4.   mandibula (f).
12.   H. B. S. Cooke, B. D. Malan and L. H. Wells 1945.
13.   H. de Villiers 1973, Human skeletal remains from Border Cave, Ingwavuma District, Kwazulu, South Africa, *Ann. Transv. Mus.*, **28** : 229–256.   (Border Cave 1, 2 and 3).   H. de Villiers 1976 (Border Cave 4).
14.   H. de Villiers 1973.   Proto-negriform, or *Homo sapiens afer* Linn.
15.   H. de Villiers 1973.   H. de Villiers 1976.
16.   P. B. Beaumont and A. K. Boshier 1972, Some comments on recent findings at Border Cave, Northern Natal, *S. Afr. J. Sci.*, **68** : 22–24.   L. H. Wells 1969, *Homo sapiens afer* Linn.   Content and earliest representatives, *S. Afr. archaeol. Bull.*, **24** : 172–173.   L. H. Wells 1972, Late Stone Age and Middle Stone Age toolmakers, *S. Afr. archaeol. Bull.*, **27** : 5–9.
17.   Department of Anatomy, University of the Witwatersrand, Johannesburg, Medical School, Hospital Street, Johannesburg.   Reg. Nos. A 1102a (Border Cave 1); A 1102b (Border Cave 2); A 1106 (Border Cave 3); A 3064 (Border Cave 4).
18.   Department of Anatomy, University of the Witwatersrand, Johannesburg, Medical School, Hospital Street, Johannesburg.

BOSKOP

1.   Boskop.
2.   Open site in cultivated field on the E bank of the Mooi River, Kolonies Plaas farm, Boskop, Potchefstroom district, SW Transvaal.   26° 34′ S, 27° 07′ E.
3.   F. W. Fitzsimons, November 1913 (calvaria); 1914.
4.   Lateritic breccia lumps at about 1·4 m in cultivated soil and disturbed sub-soil.   S. H. Haughton 1917, Preliminary Note on the Ancient Human Skull Remains from the Transvaal, *Trans. R. Soc. S. Afr.*, **6** : 1–14.
5.   ————
6.   Late Pleistocene.
7.   Middle Stone Age.   C. van Riet Lowe 1954, An Artefact Recovered with the Boskop Calvaria, *S. Afr. archaeol. Bull.*, **9** : 135–137.
8.   None reported.
9.   Skull: F = 0·2%, 100F/P205 = 1·7, eU308 = 4 ppm, N = 0·04%.   These results strongly support Pleistocene age.
10.   ————
11.   **Boskop 1.**   adult: calotte (f), temporale (f), lt ramus mandibulae (ff), lt radius (ff), lt ulna (ff), rt? femur fragments (ff), lt tibia (ff), fibula (ff).   **Holotype** of *Homo capensis* Broom, 1917.   Plate 6.
12.   F. W. Fitzsimons 1915, Palaeolithic Man in South Africa, *Nature, Lond.*, **95** : 615–616.
13.   R. Broom 1917, *in* S. H. Haughton 1917.   *Homo capensis*.   R. B. Thomson 1917, Note upon the Fragments of the Limb-bones of the Boskop Remains, appendix to S. H. Haughton 1917.   R. Broom 1918, Evidence afforded by the Boskop Skull of a New Species of Primitive Man *(Homo capensis)*, *Anthrop. Pap. Am. Mus. nat. Hist.*, **23** : 67–79.
14.   R. Singer 1958, The Boskop ' Race ' Problem, *Man,* **58** : 173–178.

H

15. S. H. Haughton 1917.
16. A. Galloway 1937, The Characteristics of the Skull of the Boskop Physical Type, *Am. J. phys. Anthrop.*, **23** : 31–46.   T. F. Dreyer *et al.* 1938, A Comparison of the Boskop with other Abnormal Skulls from South Africa, *Z. Rassenk.*, **7** : 289–296.   L. H. Wells 1959, The Problem of Middle Stone Age Man in Southern Africa, *Man,* **59** : 158–160.   P. V. Tobias 1959, The History and Metamorphosis of the Boskop Concept, *in* A. Galloway 1959, *The Skeletal Remains of Bambandyanalo,* Johannesburg : 137–146.   R. Singer 1961, Pathology in the Temporal Bone of the Boskop Skull, *S. Afr. archaeol. Bull.,* **16** : 103–104.
17. Port Elizabeth Museum, Humewood, Port Elizabeth, Cape Province.   Reg. No. PEM 1411.
18. ————

BUSHMAN ROCK SHELTER

1. Bushman Rock Shelter.
2. Rock-shelter on south-facing dolomite ridge on the main Ohrigstad-Tzaneen road to the Echo caves, Ohrigstad, Eastern Transvaal.   24° 35′ S, 30° 38′ E.
3. Mr van Zyl, 1969. (BRS 1); J. F. Eloff, 1971 (BRS 2).
4. Between layers 14 and 18, possibly layers 16/17.   R. Protsch and H. de Villiers 1974, Bushman rock shelter, Origstad, Eastern Transvaal, South Africa, *J. hum. Evol.,* **3** : 387–396.
5. ————
6. Pleistocene.
7. Earlier Late Stone Age.   R. Protsch and H. de Villiers 1974.   Middle Stone Age (Pietersburg) to Late Stone Age (Transvaal Smithfield).   J. F. Eloff 1969, Bushman rockshelter, Eastern Transvaal, *S. Afr. archaeol. Bull.,* **24** : 60.
8. *Equus burchelli, Redunca arundinum, Phacochoerus aethiopicus* cf. *Alcelaphus, Connochaetes taurinus, Aepyceros melampus.*   C. K. Brain 1969, Faunal remains from the Bushman rock shelter eastern Transvaal, *S. Afr. archaeol. Bull.,* **24** : 52–55.
9. BRS1: N = 0·86%.   Animal bones from layers 14–21: N = 0·93–0·64%.   R. Protsch and H. de Villiers 1974.
10. A3: 10,000–12,500 BP on the basis of radiocarbon dating of material from layers 12 (9940 ± 80 BP, GrN–4813), 21 (12,090 ± 95 BP, GrN–4814), 27 (12,160 ± 95 BP, GrN–4815), 28 (12,510 ± 105 BP, GrN–4816).   Middle Stone Age layers are dated > 51,000 BP (GrN–5116).   J. C. Vogel 1969, Radiocarbon dating of Bushman rockshelter, Ohrigstad District, *S. Afr. archaeol. Bull.,* **24** : 56.
    A3: 27,400–31,900 BP on basis of C14 dating of collagen in animal bones.   R. Protsch and H de Villiers 1974.
11. **BRS 1.** mandibula (f).
    BRS 2. infant 6–8 months.
12. R. Protsch and H. de Villiers 1974.
13. R. Protsch and H. de Villiers 1974.
14. ————
15. R. Protsch and H. de Villiers 1974.
16. ————
17. Department of Archaeology, University of Pretoria, Pretoria.
18. ————

BYNESKRANSKOP　　　　　　　34° 35′ S, 19° 28′ E.

In April 1973, F. Schweitzer excavated a juvenile skeleton from a burial mound in a limestone cave, 12 km E of Die Kelders, 200 m above sea level near Walker Bay, Cape Province.　It was associated with a Late Stone Age industry.　Charcoal indicated a C14 date between 5000 and 10,500 BP (UW).

The remains are preserved in the South African Museum, Cape Town.

CAPE FLATS

1.　Cape Flats (Philippi).
2.　Sand pit on the Cape Flats, near Cape Town.　34° 25′ S, 18° 27′ E.
3.　M. R. Drennan, 1929.
4.　Old land surface at 1 m within wind-blown sand.　M. R. Drennan 1929, An Australoid Skull from the Cape Flats, *Jl R. anthrop. Inst.*, **59** : 417–427.
5.　————
6.　Late Pleistocene/Holocene.
7.　Late Middle Stone Age or 2nd Intermediate (Magosian).　A. J. H. Goodwin 1929, Report on the Stone Implements found with the Cape Flats Skull, *Jl R. anthrop. Inst.*, **59** : 429–438.
8.　None reported.
9.　Cape Flats skull: F = 0·1%, 100F/P205 = 0·15, N = 1·7%.　Inferred Holocene age.　K. P. Oakley 1957, The Dating of the Broken Hill, Florisbad and Saldanha Skulls, *Proc. 3rd Pan-Afr. Congr. Prehist.* : 76–79.　L. H. Wells 1959, The Problem of Middle Stone Age Man in Southern Africa, *Man,* **59** : 158–160.
10.　————
11.　**Cape Flats 1.**　adult: cranium (f), mandibula (f), femur (i).　**Holotype** of *Homo drennani* Kleinschmidt, 1931.　Plate 7.
　　　Cape Flats 2.　adult: cranium (ff).
　　　Fragments of limb-bones belonging to at least three different individuals.
12.　M. R. Drennan 1929 (Cape Flats 1).　*Homo australoideus africanus.*　(Cape Flats 2).　Bush-Hottentot.
13.　M. R. Drennan 1929.
14.　————
15.　M. R. Drennan 1929.
16.　A. Galloway 1937, Man in Africa in the Light of Recent Discoveries, *S. Afr. J. Sci.,* **34** : 89–120.
17.　Department of Anatomy, University of Cape Town, P.O. Box 594, Rondebosch, Cape Province.　Reg. No. UCT 62.
18.　————

CAVE OF HEARTHS

1. Cave of Hearths.
2. Rock-shelter in Makapansgat valley, 16 km ENE of Potgietersrus, Transvaal.   24° 12′ S, 28° 57′ E.
3. B. Kitching, September 1947.
4. Cave breccia, Bed 3, calcreted red earth.   R. J. Mason 1962, *Prehistory of the Transvaal,* Johannesburg.
5. ─────
6. Upper Pleistocene.   R. J. Mason 1962.
7. Final Acheulian/Fauresmith.   J. D. Clark 1964, The Later Pleistocene Culture of Africa, *Science, N.Y.,* **150** : 838.   R. J. Mason, *Cave of Hearths in Prehistory,* Johannesburg.
8. *Equus helmei, Homoioceras baini, Alcelaphus helmei.*   H. B. S. Cooke 1962, Notes on the Faunal Material from the Cave of Hearths and Kalkbank, Appendix 1, *in* R. J. Mason 1962 : 447–453.   H. B. S. Cooke 1963, Pleistocene Mammal Faunas of Africa, with particular Reference to Southern Africa, *in* F. C. Howell and F. Bourlière (eds) 1963, African Ecology and Human Evolution, *Publs Anthrop. Viking Fund,* **36** : 91–95.
9. Human mandible: $F = 1·0\%$, $100F/P205 = 3·5$, $N = 0·17\%$.   Animal bone: $F = 1·2\%$, $100F/P205 = 3·4$, $eU308 = 2$ ppm, $N = 0·7\%$.   Another animal bone: $N = 0·06\%$ (UCLA–1752A).
10. A3: *c.* 55,000 BP on basis of C14 dating of wood of Final Acheulian/Fauresmith stage at Kalambo Falls: > 52,000 BP (GrN–1396).   J. D. Clark 1965: 836.   Only available C14 dating at site is based on samples from overlying Middle Stone Age layers and ranged up to $16,811 \pm 960$ BP (C–926).
11. **Cave of Hearths 1.**   adolescent about 12 yrs: rt corpus mandibulae, 2I1, I2, P3, M1, M2; radii (ff).
12. R. A. Dart 1948, The First Human Mandible from the Cave of Hearths, Makapansgat, *S. Afr. archaeol. Bull.,* **3** : 96–98.   Neandertaloid, cf. Boskop.
13. P. V. Tobias 1968, Middle and Early Upper Pleistocene Members of the Genus *Homo* in Africa, *in* G. Kurth (ed.) 1968, *Evolution und Hominisation,* Stuttgart : 183–184.
14. P. V. Tobias 1971, Human Skeletal Remains from the Cave of Hearths, Makapansgat, Northern Transvaal, *Am. J. phys. Anthrop.* **34** : 335–367.
15. R. A. Dart 1948.
16. L. H. Wells 1959, *in* J. D. Clark 1959, *The Prehistory of Southern Africa,* London.   P. V. Tobias 1961, New Evidence and New Views on the Evolution of Man in Africa, *S. Afr. J. Sci.,* **57** : 25–38.
17. Department of Anatomy, University of the Witwatersrand, Medical School, Hospital Street, Johannesburg, Transvaal.   Field Reg. No. D5/10.
18. Department of Anatomy, University of the Witwatersrand, Medical School, Hospital Street, Johannesburg, Transvaal.

ELANDSFONTEIN

See under **HOPEFIELD**

FISH HOEK

1. Fish Hoek (Skildergat).
2. Rock-shelter, Skildergat or Peers' Cave, Fish Hoek, about 23 km S of Cape Town, Cape Province.   34° 20′ S, 18° 24′ E.
3. V. S. and B. Peers, 1927–29.
4. Cave deposit.   A. Keith 1931, *New Discoveries Relating to the Antiquity of Man,* London: 126–142.
5. Contemporaneous burial.   A. Keith 1931.
6. Pleistocene.
7. Stillbay/Howieson's Poort.   L. H. Wells 1959, The problem of Middle Stone Age man in Southern Africa, *Man,* **59** : 158–160.   R. J. Mason 1962, *Prehistory of the Transvaal,* Johannesburg : 233.   R. G. Klein 1970, Problems in the study of the Middle Stone Age of South Africa, *S. Afr. archaeol. Bull.,* **25** : 127–135.
8. *Equus capensis.*   R. A. Dart 1933, Fossil Man and Contemporary Faunas in Southern Africa, *Int. geol. Congr.* : 1249–1270.
9. Fish Hoek skull : F = 0·4%, 100F/P205 = 1·4, N = 0·6%.
    Animal bone Howieson's Poort (= Magosian) level : F = 0·2%, 100F/P205 = 0·5, N = 0·6%.
    Shell mound skeleton : F = 0·1%, 100F/P205 = 0·4, N = 4·6%.
    Inferred age: Magosian.   K. P. Oakley 1957, The Dating of the Broken Hill, Florisbad and Saldanha Skulls, *Proc. 3rd Pan-Afr. Congr. Prehist.* : 76–79.
10. A2: > 35,600 BP (GX–267) on basis of C14 dating of sample of charcoal from 180 cm below datum, associated with ? Stillbay assemblage.   *Radiocarbon,* **8** : 158.
    A2: 36,000 ± 2400 BP (UCLA–1235) on basis of C14 dating of charcoal from 237–242 cm below datum, at base of hearth area, in square A–2 of trench 2, associated with a ? Stillbay assemblage.   B. W. Anthony 1967, Excavations at Peers Cave, Fish Hoek, South Africa, *in* E. M. van Zinderen Bakker (ed.) 1967, *Palaeoecology of Africa,* Vol. II, Cape Town : 58–59.   R. G. Klein 1970.   P. B. Beaumont and J. C. Vogel 1972, On a new radiocarbon chronology for Africa south of the Equator, Part II, *African Studies* : 155–182.
11. **Fish Hoek 1.**   adult: cranium (i), mandibula (i), post-cranial skeleton (i).
12. A. Keith 1931.
13. A. Keith 1931.   Bushmanoid.
14. ————
15. A. Keith 1931.
16. R. A. Dart 1940, Recent Discoveries Bearing on Human History in Southern Africa, *Jl. R. anthrop. Inst.,* **70** : 13–27.   K. Jolly 1947, Preliminary Note on New Excavations at Skildergat, Fish Hoek, *S. Afr. archaeol. Bull.,* **2** : 11–12.   K. Jolly 1948, The Development of the Cape Middle Stone Age in the Skildergat cave, Fish Hoek, *S. Afr. archaeol. Bull.,* **3** : 106–107.   P. V. Tobias 1971, Human Skeletal Remains from the Cave of Hearths, Makapansgat, Northern Transvaal, *Am. J. phys. Anthrop.,* **34** : 335–367.
17. South African Museum, P.O. Box 61, Cape Town, Cape Province, Reg. No. SAM 4692.
18. ————

FLORISBAD

1. Florisbad.
2. Open site, 40 km NW of Bloemfontein, Orange Free State.   28° 45′ S, 26° 00′ E.
3. T. F. Dreyer, 1932.
4. Eye of a fossil spring.   K. P. Oakley 1954, Study Tour of Early Hominid Sites in Southern Africa, *S. Afr. archaeol. Bull.,* **9** : 75–87.
5. ————
6. Upper Pleistocene, Florisbad-Vlakkraal stage.   H. B. S. Cooke 1963, Pleistocene Mammal Faunas of Africa, with particular reference to Southern Africa, *in* F. C. Howell and F. Bourlière (eds) 1963, African Ecology and Human Evolution, *Publs Anthrop. Viking Fund,* **36** : 89.
7. Middle Stone Age, Hagenstadt variant.   K. P. Oakley 1954.
8. *Hyaena brunnea, Equus helmei, Equus burchelli, Diceros bicornis, Phacochoerus aethiopicus, Hippopotamus amphibius, Homoioceras baini, Kobus venterae, Alcelaphus helmei, Gazella bondi.*   H. B. S. Cooke 1963 : 98–101.
   Peat I. E. M. van Zinderen Bakker 1957, A Pollen Analytical Investigation of the Florisbad Deposits (South Africa), *Proc. 3rd Pan-Afr. Congr. Prehist.* : 56–57.
9. Florisbad human nasalia: F = 1·6%.   100F/P205 = 6·6, N = 0·4%; maxilla: F = 1·0%, 100F/P205 = 9·5, N = 1·7%.   Bone from peat I: F = 2·1%, 100F/P205 = 7·4, N = nil.   Bone from ' spring eye ': F = 2·8%, 100F/P205 = 9·0, N = 0·6%.   K. P. Oakley, 1957, The Dating of the Broken Hill, Florisbad and Saldanha Skulls, *Proc. 3rd Pan-Afr. Congr. Prehist.* : 76–79.
10. A2: 38,500 BP on basis of C14 dating of wood and collagen samples (UCLA–1745B, C).   J. L. Bada, R. Protsch and R. A. Schroeder 1973, The Racemization reaction of Isoleucine used as a Palaeotemperature Indicator, *Nature, Lond.,* **241** : 394–395. > 44,000 BP (Y–103) based on C14 dating of Peat I soil samples.   G. W. Barendsen, E. S. Deevey and L. J. Gralenski 1957, Yale natural radiocarbon measurements III, *Science,* **126** : 908.   > 48,900 BP (GrN–4208) on basis of C14 dating of Peat I. J. C. Vogel and P. B. Beaumont 1972, Revised Radiocarbon Chronology for the Stone Age in South Africa, *Nature, Lond.,* **237** : 50–51.
11. **Florisbad 1.**   adult: cranium (ff).   **Holotype** of *Homo (Africanthropus) helmei* Dreyer, 1935.   Plate 7.
12. T. F. Dreyer 1935, A Human Skull from Florisbad, Orange Free State, with a note on the endocranial cast, by C. U. Ariens Kappers, *Proc. Acad. Sci. Amst.,* **38** : 119–128.
13. T. F. Dreyer 1935.
14. H. V. Vallois 1957, *Fossil Men,* New York: 1–535.   *Homo sapiens.*
15. T. F. Dreyer 1935.
16. M. R. Drennan 1937, The Florisbad Skull and Brain Cast, *Trans. R. Soc. S. Afr.,* **25** : 103–114.   A. Galloway 1937, The Nature and Status of the Florisbad Skull as revealed by its non-metrical Features, *Am. J. phys. Anthrop.,* **23** : 1–16.   L. H. Wells 1947, Human Crania of the Middle Stone Age in South Africa, *Proc. 1st Pan-Afr. Congr. Prehist.* : 125–133.   L. H. Wells 1959, The Problem of Middle Stone Age Man in Southern Africa, *Man,* **59** : 158–160.   P. V. Tobias 1968, Middle and Early Upper Pleistocene Members of the genus *Homo* in Africa, *in* G. Kurth (ed.) 1968, *Evolution und Hominisation* (2nd edn) Stuttgart : 176–194.
17. National Museum, Bloemfontein, Orange Free State.
18. National Museum, Bloemfontein, Orange Free State.

HOPEFIELD

1. Hopefield (Saldanha).
2. Open site, Elandsfontein farm, near Hopefield, 24 km E of Saldanha Bay, Cape Province.   33° 04′ S, 18° 23′ E.
3. K. Jolly and R. Singer, 8 January 1953.
4. Surface of sand and calcareous concretions.   R. Singer 1954, The Saldanha Skull from Hopefield, South Africa, *Am. J. phys. Anthrop.*, **12** : 345–362.
5. —————
6. Upper Pleistocene, Vaal-Cornelia stage.   H. B. S. Cooke 1963, Pleistocene Mammal Faunas of Africa, with Particular Reference to Southern Africa, *in* F. C. Howell and F. Bourlière (eds) 1963, African Ecology and Human Evolution, *Publs Anthrop. Viking Fund,* **36** : 89.
7. Late South African Acheulian.   J. J. Wymer 1973, *in lit.*   R. Singer 1956, *Am. Anthrop.* **58** : 1127–1134.
8. *Mesochoerus lategani, Equus* cf. *plicatus, Loxodonta* cf. *atlantica, Hippopotamus amphibius, Homoioceras* cf. *baini.*   H. B. S. Cooke 1963.   Early-Middle Pleistocene fauna.   R. Klein 1973, Geological antiquity of Rhodesian Man, *Nature, Lond.,* **244** : 311–312.
9. Saldanha hominid calvaria: $F = 2 \cdot 0\%$, $100F/P205 = 5 \cdot 9$, $eU308 = 21$ ppm, $N = 0 \cdot 2\%$.   Animal bones from main fossil bed: $F = 1 \cdot 7–2 \cdot 5\%$, $100F/P205 = 5 \cdot 2–7 \cdot 8$, $eU308 = 6–25$ ppm, $N = 0 \cdot 1–0 \cdot 5\%$.   K. P. Oakley 1957, The Dating of the Broken Hill, Florisbad and Saldanha Skulls, *Proc. 3rd Pan-Afr. Congr. Prehist.* : 76–79.
10. —————
11. **Hopefield 1.**   adult: calvaria (f), rt ascending ramus mandibulae (f).   **Holotype** of *Homo saldanensis* Drennan, 1955.   Plate 8.
12. M. R. Drennan 1953, The Saldanha Skull and its Associations, *Nature, Lond.,* **172** : 791–793.
13. R. Singer 1954.   Cf. Broken Hill Skull.   M. R. Drennan and R. Singer 1955, A mandibular fragment, probably of the Saldanha skull, *Nature, Lond.,* **175** : 364.
14. —————
15. R. Singer 1954.
16. R. Singer and J. R. Crawford 1958, The Significance of the Archaeological Discoveries at Hopefield, South Africa, *J. R. anthrop. Inst.,* **88** : 11–19.   F. C. Howell and J. D. Clark 1963 : 474–476.   *Homo sapiens rhodesiensis.*   R. Singer 1957, Recent discoveries in Southern Africa, *Mitt. anthrop. Ges. Wien,* **87** : 110.   Cf. Ngandong.
17. South African Museum, P.O. Box 61, Cape Town, Cape Province.
18. —————

KLASIES RIVER MOUTH

1. Klasies River Mouth.
2. Cave No. 1, 40 km W of Cape St Francis and 120 km W of Port Elizabeth, Cape Province.   34° 04′ S, 24° 12′ E.
3. Team working with J. J. Wymer, University of Chicago Archaeological Expedition, directed by R. Singer, about 1968.

4. Cave deposits, layer 14.
5. No evidence of burial.
6. Upper Pleistocene.   Last Interglacial.
7. Full Middle Stone Age, Mossel Bay phase (Eds).   R. Singer and J. Wymer 1969, A new Middle Stone Age cave site in South Africa, *Am. J. phys. Anthrop.,* **31** : 256.   Late Stone Age.   C. G. Sampson 1968, The Middle Stone Age industries of the Orange River Scheme area, *Nasional Mus. Bloemfontein,* Mem. 4 : 110.
8. Fauna mainly extant forms, but some extinct, including *Pelorovis*.   R. G. Klein 1974, Environment and subsistence of prehistoric man in the Southern Cape Province, South Africa, *Wld Archaeol.,* **5** : 249–284.
9. KRM 1: N = 0·26%; KRM 2: N = 0·20% (UCLA–1743).
10. A3: > 38,000 BP on basis of C14 dating of carbon in overlying layer of Final Middle Stone Age (Magosian = Howieson's Poort phase).   R. G. Klein 1974.
11. **KRM 1.**   adult: mandibula with dentes.
    KRM 2.   adult: mandibula with dentes.
    KRM unnumbered.*
12. R. Singer and P. Smith 1969, Some human remains associated with the Middle Stone Age deposits at Klasies River, South Africa, *Am. J. phys. Anthrop.,* **31** : 256.   KRM 1 and 2 represent two populations.
13. R. Singer *et al.* in preparation.
14. ————
15. ————
16. J. J. Wymer and R. Singer 1972, Middle Stone Age occupational settlements on the Tzitzikama coast, eastern Cape Province, South Africa, *in* P. J. Ucko, R. Tringham and G. W. Dimbleby (eds) 1972, *Man, Settlement and Urbanism,* London : 207–210.   R. G. Klein 1975, Middle Stone Age man-animal relationships in Southern Africa: Evidence from Die Kelders and Klasies River Mouth, *Science, N.Y.,* **190** : 265–267.
17. South African Museum, Cape Town.   Temporarily in Department of Anatomy, University of Chicago, Chicago XX, Illinois 60637, U.S.A.
18. ————

KLIPFONTEINRAND   32° 04′ S, 19° 08′ E.

Skeletons of an adult and infant were excavated by J. E. Parkington on 5 December 1969 from the second rock shelter at Klipfonteinrand, Clanwilliam District, 240 km N of Cape Town.   They were crouched burials associated with Late Stone Age artifacts, 20–30 cm below the surface.

J. E. Parkington and C. Poggenpoel 1971, A Late Stone Age Burial from Clanwilliam, *S. Afr. archaeol. Bull.,* **26** : 82–84.

The remains are preserved in the Department of Archaeology, University of Cape Town.

---

*Numerous, unpublished, fragmentary hominid skeletal materials found during excavations in the Klasies River Mouth Caves are preserved in the South African Museum, Cape Town (*fide* T. H. Barry).

KROMDRAAI

1. Kromdraai (1938 cranium).
2. Breccia-filled hollow in Kromdraai kopje, 1·6 km ENE of Sterkfontein, 9·6 km N of Krugersdorp, Transvaal.   25° 59′ S, 27° 47′ E.
3. G. Terblanche, 8 June 1938 (TM 1517); R. Broom, 1944 (TM 1603).
4. Phase II Breccia, upper 3 m at Site B.   C. K. Brain 1958, The Transvaal Ape-Man-Bearing Cave Deposits, *Transv. Mus. Mem.*, **11** : 95–99.
5. ————
6. Early Middle Pleistocene (youngest of the South African australopithecine deposits).   K. P. Oakley 1954, Dating of the Australopithecinae of Africa, *Am. J. phys. Anthrop.*, **12** : 16–17.
7. One flake artifact.   C. K. Brain 1958 : 98.
8. Cornelia span with cf. *Connochaetes taurinus, Antidorcas* cf. *recki, A. bondi, Gazella, Papio robinsoni, P. Angusticeps, Cercopithecoides williamsoni.*   L. Freedman and C. K. Brain 1972, Fossil cercopithecoid remains from the Kromdraai Australopithecine site (Mammalia: Primates), *Ann. Transv. Mus.*, **28** : 1–16.   E. S. Vrba 1975, Some Evidence of Chronology and Palaeoecology of Sterkfontein, Swartkrans and Kromdraai from the fossil Bovidae, *Nature, Lond.*, **254** : 301–304.   E. S. Vrba 1976, The Fossil Bovidae of Sterkfontein, Swartkrans and Kromdraai, *Transv. Mus. Mem.*, **21** : 1–166.
9. Hominid bone TM 1517: F = 0·3%, 100F/P205 = 1·3, eU308 = 5 ppm, N = nil.   Animal bone Site A: F = 0·7%, 100F/P205 = 1·7, eU308 = 3 ppm, N = nil.
10. ————
11. **TM 1517.**   adult, probably male: lt part of cranium, including parietale (f), temporale and ossa faciei, palatinum, lt maxilla with P3, P4, M1, M2; isolated crowns of upper rt P3, P4, M1, M2, M3; rt mandibula (f) with P3, P4, M1, M2, M3 and cast of rt C; isolated lt P3, P4.   Associated proximal end of rt ulna, distal end of rt. humerus, lt metacarpale II (f), rt talus (f), proximal phalanx manus, proximal phalanx and distal phalanx pedis.   **Holotype** of *Paranthropus robustus* Broom, 1938.   Plate 9.
    TM 1603.   upper lt M3 crown from same individual.
12. R. Broom 1938a, The Pleistocene Anthropoid Apes of South Africa, *Nature, Lond.*, **142** : 377–379.   R. Broom 1938b, Further Evidence on the Structure of the South African Pleistocene Anthropoids, *Nature, Lond.*, **142** : 897–899.
13. R. Broom and G. W. H. Schepers 1946, The South African Fossil Ape Men : the Australopithecinae, *Transv. Mus. Mem.*, **2** : 84–109, 113–118. *Paranthropus robustus* holotype.   W. K. Gregory and M. Hellman 1939, The dentition of the extinct South African Man-Ape *Australopithecus (Plesianthropus) transvaalensis* Broom, *Ann. Transv. Mus.*, **19** : 339–373.   R. Broom 1942, The hand of the Ape-Man, *Paranthropus robustus, Nature, Lond.*, **149** : 513.   R. Broom 1943, An ankle-bone of the Ape-Man *Paranthropus robustus, Nature, Lond.*, **152** : 689.   J. T. Robinson 1956, The Dentition of the Australopithecinae, *Transv. Mus. Mem.*, **9** : 57, 62, 71, 75, 83, 88, 92, 104, 109, 113.
14. J. T. Robinson 1972, *Early Hominid Posture and Locomotion:* 1–361, Chicago and London (post-cranial bones).
15. R. Broom and G. W. H. Schepers 1946, pls 8, 9, 10 (figs 100–113); pl. 11 (figs 134–136, 139); pl. 12 (figs 142–147); pl. 13 (figs 157–160, 166, 170, 173, 176, 179,

182).   J. T. Robinson 1972, figs 86–7, 92–100, 105.

16. M. Day and B. Wood 1968, Functional affinities of the Olduvai Hominid 8 talus, *Man*, **3** : 440–455.   W. E. Le Gros Clark 1947, Observations on the Anatomy of the Fossil Australopithecinae, *J. Anat.*, **81** : 319–324.   W. L. Straus 1948, The Humerus of *Paranthropus robustus*, *Am. J. phys. Anthrop.*, **6** : 285–311.   S. L. Washburn and B. Patterson 1951, Evolutionary importance of the South African man-apes, *Nature, Lond.*, **167** : 650.   *Australopithecus* sp.   J. T. Robinson 1954, The Genera and Species of the Australopithecinae, *Am. J. phys. Anthrop.*, **12** : 181–200.   *Paranthropus robustus robustus*.   H. M. McHenry 1974, How large were the Australopithecines?, *Am. J. phys. Anthrop.*, **40** : 329–340 (TM 1517).   W. E. Le Gros Clark 1967, *Man-Apes or Ape-Men?*, New York, 1–150.   *Australopithecus robustus*.

17. Transvaal Museum, P.O. Box 413, Pretoria, Transvaal.

18. Wenner-Gren Foundation for Anthropological Research, 14 East 71st Street, New York, N.Y. 10021, U.S.A. (TM 1517: mandible, distal humerus, proximal ulna, talus).

1. Kromdraai (1938–1955).

2. Breccia-filled hollow in Kromdraai kopje, 1·6 km ENE of Sterkfontein, 9·6 km N of Krugersdorp, Transvaal.   25° 59′ S, 27° 47′ E.

3. A. L. Du Toit, February 1941 (TM 1536).   R. Broom, 1938? (TM 1602).   C. K. Brain and B. J. Grobbelaar, 1954 (TM 1600, TM 1601, TM 1605); 1955 (TM 1604).

4. Phase II Breccia, upper 3 m at site B.   C. K. Brain 1958, The Transvaal Ape-Man-Bearing Cave Deposits, *Transv. Mus. Mem.*, **11** : 95–99.

5. ———

6. Early Middle Pleistocene (youngest of the South African australopithecine deposits).   K. P. Oakley 1954, The Dating of the Australopithecinae of Africa, *Am. J. phys. Anthrop.*, **12** : 16–17.

7. One flake artifact.   C. K. Brain 1958 : 98.

8. Cornelia span with cf. *Connochaetes taurinus*, *Antidorcas* cf. *recki*, *A. bondi*, *Gazella*.   E. S. Vrba 1975, Some evidence of chronology and palaeoecology of Sterkfontein, Swartkrans and Kromdraai from the fossil Bovidae, *Nature, Lond.*, **254** : 301–304.

9. ———

10. ———

11. **TM 1536.**   juvenile, possibly female: mandibula (f) with rt I1, di2, dc, dm1, dm2, M1; lt dc (f), dm1.
    TM 1600.   mandibula (ff) with lt P M2, M3.
    TM 1601.   isolated teeth, lower dm1, crowns of C, P3, P4; upper M1.
    TM 1602.   rt maxilla (ff) with roots of M1, M2, M3; palatinum (ff).
    TM 1604.   isolated lower dm2 (f) without roots.
    TM 1605.   ilium, including part of acetabulum, lacking iliac crest.

12. R. Broom 1941, Mandible of a young *Paranthropus* child, *Nature, Lond.*, **147** : 607 (TM 1536).   Unpublished: TM 1600, TM 1601, TM 1602, TM 1604, TM 1605.

13. R. Broom and G. W. H. Schepers 1946, The South African Fossil Ape-Men: the Australopithecinae, *Transv. Mus. Mem.*, **2** : 106, 109–113 (TM 1536).   J. T. Robinson 1956, The Dentition of the Australopithecinae, *Transv. Mus. Mem.*, **9** : 130, 132, 136, 141 (TM 1536).
14. J. T. Robinson 1972, *Early Hominid Posture and Locomotion,* Chicago and London : 88–91 (TM 1605).
15. R. Broom and G. W. H. Schepers 1946, pls 8, 9, 10 (figs 100–113); pl. 11 (figs 134, 136, 139); pl. 12 (figs 142–147); pl. 13 (figs 157–160, 166, 170, 173, 176, 179, 182).   J. T. Robinson 1972, fig. 49 (TM 1605).
16. W. E. Le Gros Clark 1947, Observations on the Anatomy of the Fossil Australopithecinae, *J. Anat.*, **81** : 319–324.   W. L. Straus 1948, The Humerus of *Paranthropus robustus, Am. J. phys. Anthrop.*, **6** : 285–311.   J. T. Robinson 1954, The Genera and Species of the Australopithecinae, *Am. J. phys. Anthrop.*, **12** : 181–200.   *Paranthropus robustus robustus.*
17. Transvaal Museum, P.O. Box 413, Pretoria, Transvaal.
18. Wenner-Gren Foundation for Anthropological Research, 14 East 71st Street, New York, N.Y. 10021 U.S.A. (TM 1536, TM 1605).

MAKAPANSGAT : CAVE OF HEARTHS

See under **CAVE OF HEARTHS**

MAKAPANSGAT LIMEWORKS

General Entry

1. Makapansgat.
2. Breccia filling of ancient dolomitic limestone cave, Makapan valley, 16 km ENE of Potgietersrus, Transvaal.   24° 12′ S, 28° 57′ E.
3. J. Kitching, A. R. Hughes and the Staff of the Department of Anatomy of the University of the Witwatersrand under the direction of Professor R. A. Dart, 1947–1961.   J. Kitching, 1975.
4. Some specimens were found in undisturbed breccia deposits, but the majority were found in the breccia dumps of the travertine miners.   Two main phases of breccia deposits have been recognized both in undisturbed and mined breccia.   The hominids are all from Upper Phase I (Member 4) and Lower Phase I (Member 3).   The ' cone ' is a large lump of breccia which has fallen from the roof of the modern cave.   C. K. Brain 1958, The Transvaal Ape-Man-Bearing Cave Deposits, *Transv. Mus. Mem.*, **11** : 107–111.   K. W. Butzer 1971, Another look at the australopithecine cave breccias of the Transvaal, *Am. Anthrop.*, **73** : 1197–1201.
5. ————

6. ' Late Villafranchian. ' K. P. Oakley 1957, Dating the Australopithecines, *Proc. 3rd Pan-Afr. Congr. Prehist.* : 155–157. First Interpluvial. C. K. Brain 1958 : 118. ' Upper Villafranchian. ' H. B. S. Cooke 1963, Pleistocene Mammal Faunas of Africa, *in* F. C. Howell and F. Bourlière (eds) 1963, African Ecology and Human Evolution, *Publs Anthrop. Viking Fund.* **36** : 103. ' Sterkfontein Faunal Span .' H. B. S. Cooke 1968, Evolution of Mammals on Southern Continents. II. The Fossil Mammals of Africa, *Quart. Rev. Biol.,* **43** (3) : 234–264. L. H. Wells 1969, Faunal subdivision of the Quaternary in southern Africa, *S. Afr. archaeol. Bull.,* **24** : 93–95.

7. Pebble Culture. R. A. Dart 1955, The First Australopithecine Fragment from the Makapansgat Pebble Culture Stratum, *Nature, Lond.,* **176** : 170–171. Utilized bones? R. A. Dart 1957, The Makapansgat Australopithecine Osteodontokeratic Culture, *Proc. 3rd Pan-Afr. Congr. Prehist.* : 161–171. R. A. Dart 1957, The Osteodontokeratic Culture of *Australopithecus prometheus, Transv. Mus. Mem.,* No. **10**. R. A. Dart 1962, Substitution of stone tools for bone tools at Makapansgat, *Nature, Lond.,* **196** : 314–316. B. Maguire 1965, Foreign pebble pounding artefacts in the breccias and the overlying vegetation soil at Makapansgat Limeworks, *S. Afr. archaeol. Bull.,* **20** : 117–130. P. V. Tobias 1968, Cultural hominisation among the earlier African Pleistocene hominids, *Proc. Prehist. Soc.,* n.s. **33** : 367–376. D. L. Wolberg 1970, The Hypothesized Osteodontokeratic Culture of the Australopithecinae: A look at the Evidence and Opinions, *Curr. Anthrop.,* **11** : 23–37.

8. Sterkfontein Faunal Span, including *Parapapio broomi, P. whitei, P. jonesi, Cercopithecoides williamsi, Simopithecus darti, Dinofelis barlowi, Procavia transvaalensis, P. antiqua, Equus helmei, Makapania broomi, Redunca darti, Xenohystrix* sp., *H. hyaena* ssp., *Gigantohyrax, Hipparion* cf. *libycum, Potamochoeroides, Notochoerus, Libytherium.* R. F. Ewer 1957, Faunal evidence on the dating of the Australopithecinae, *Proc. 3rd Pan-Afr. Congr. Prehist.,* 136–137. L. H. Wells 1962, Pleistocene faunas and the distribution of mammals in southern Africa, *Ann. Cap. Prov. Museum,* II : 37–40. L. H. Wells 1967, Antelopes in the Pleistocene of Southern Africa, *in* W. W. Bishop and J. D. Clark (eds) 1967, *Background to Evolution in Africa,* Chicago and London : 99–107. H. B. S. Cooke 1963. H. B. S. Cooke 1968, Possibly earlier than Sterkfontein. L. H. Wells 1969. L. H. Wells 1971, Africa and the Ancestry of Man, *S. Afr. J. Sci.,* **67** (4) : 276–283. P. M. Butler and M. Greenwood 1976, *in* R. J. G. Savage and S. C. Coryndon (eds) 1976, Elephant Shrews (Macroscelididae) from Olduvai and Makapansgat, *Fossil Vertebrates of Africa,* Vol. 4 : 1–56.

9. ─────────

10. T. C. Partridge 1973, Geomorphological Dating of Cave Openings at Makapansgat, Sterkfontein, Swartkrans and Taung, *Nature, Lond.,* **246** : 75–79.

11.–15. See under individual entries.

16. J. T. Robinson 1954, The Genera and Species of the Australopithecinae, *Am. J. phys. Anthrop.,* **12** : 181–200. *Australopithecus africanus transvaalensis.* R. A. Dart and E. L. Boné 1955, A Catalogue of the Australopithecine Fossils found at Limeworks, Makapansgat, *Am. J. phys. Anthrop.,* **13** : 623. W. E. Le Gros Clark 1964, *The Fossil Evidence for Human Evolution,* 2nd edn. Chicago : 123–173. *Australopithecus africanus.* J. T. Robinson 1972, *Early Hominid Posture and Locomotion,* Chicago and London. *Homo africanus.* P. V. Tobias 1973, Implications of the New Age Estimates of the Early South African Hominids, *Nature, Lond.,* **246** : 79–83. H. M. McHenry 1974, How large were the Australopithecines?, *Am. J. phys. Anthrop.,* **40** : 329–340

(MLD 15, 16, 32). Le Gros Clark 1967, *Man Apes or Ape-Men?,* New York, 1–150. J. T. Robinson 1967, Variation and the taxonomy of the early hominids, *Evolutionary Biology*, Vol. I: 69–100.

17.   Department of Anatomy, University of the Witwatersrand, Johannesburg, Medical School, Hospital Street, Johannesburg, Transvaal.

18.   While stocks last: Wenner-Gren Foundation for Anthropological Research, 14 East 71st Street, New York, N.Y. 10021, U.S.A. (MLD 1, MLD 2, MLD 10, MLD 19, MLD 20).

Index of Fossil finds from Makapansgat: Limeworks

Reg. No.	Element	Page
MLD 1	adult: calvaria	110
MLD 2	mandibula	112
MLD 3	rt parietale	110
MLD 3a	endocast of MLD 3	110
MLD 4	M3	112
MLD 5	dm2	112
MLD 6	maxilla	110
MLD 7	ilium	113
MLD 8	rt ischial fragment	113
MLD 9/12	maxilla, palatinum	111
MLD 9a	breccia cast of MLD 9	111
MLD 10	biparieto-occipitale	110
MLD 10a	breccia cast of MLD 10	110
MLD 11/30	I2, C, P3, P4	111
MLD 13	C	111
MLD 14	humerus	113
MLD 15	radius	113
MLD 16	radius	113
MLD 17	femur	113
MLD 18	mandibula	112
MLD 19	mandibula	112
MLD 20	clavicula	113
MLD 20a	breccia cast of MLD 20	113
MLD 21	ilium	113
MLD 21a	breccia cast of MLD 21	113
MLD 22	mandibula	112
MLD 23	maxilla	110
MLD 23a	breccia cast of MLD 23	110
MLD 24	M2	112
MLD 24a	breccia cast of MLD 24	112
MLD 25	ilium	113
MLD 26	frontale	110
MLD 27	mandibula	112
MLD 28	maxilla, palatinum	111
MLD 29	mandibula 1956	112
MLD 31	temporale	110
MLD 32	radius	113
MLD 33	parietale	110

Reg. No.	*Element*	*Page*
MLD 34	mandibula	112
MLD 35	ilium	113
MLD 36	clavicula	111
MLD 37/38	cranium	111
MLD 39	humerus	113
MLD 40	mandibula	112

Note: The hominid status of MLD 13, MLD 14, MLD 15, MLD 17, MLD 20, MLD 21, MLD 26, MLD 31, MLD 33, MLD 36, MLD 39 is doubtful.

1. Makapansgat Limeworks (1947 calvaria).
3. J. Kitching, September 1947.
4. Limeworks dumps from Lower Phase I Breccia.   C. K. Brain 1958, The Transvaal Ape-Man-Bearing Cave Deposits, *Transv. Mus. Mem.,* **11** : 107–111.
11. **MLD 1.**   adult, probably female: calvaria (ff) including posterior parts of parietalia and most of occipitale.   **Holotype** of *Australopithecus prometheus* Dart, 1948. Plate 12.
12. R. A. Dart 1948, The Makapansgat Proto-Human *Australopithecus prometheus, Am. J. phys. Anthrop.,* **6** : 259–284.
13. R. A. Dart 1948 : 259–284.   *Australopithecus prometheus.*
14. P. V. Tobias 1967, *Olduvai Gorge,* vol. II, Cambridge.   *A. africanus* with some *A. robustus* features.   J. T. Robinson 1954a, The genera and species of the Australopithecinae, *Am. J. phys. Anthrop.,* **12** : 181–200.   *A. africanus transvaalensis.*
15. R. A. Dart 1948.
16. J. T. Robinson 1954b, The australopithecine occiput, *Nature, Lond.,* **174** : 262.

1. Makapansgat Limeworks (1948–1956 cranial fragments).
3. A. R. Hughes, September 1948 (MLD 3).   B. Kitching, October 1948 (MLD 6), November 1948 (MLD 10).   A. R. Hughes, 16 August 1955 (MLD 23), 24 July 1956 (MLD 26).   J. Kitching, October–December 1956 (MLD 31, MLD 33).
4. Limeworks dumps from Lower Phase I Breccia.
11. **MLD 3.**   infant: rt parietale.   MLD 3a endocast of MLD 3, including endocast of missing part near lambda.
    MLD 6.   adult female: cranio-facial fragment, including rt maxilla with rt P3 (f), P4, M1, M2.   Same individual as MLD 23 and possibly same individual as MLD 1.   MLD 6a impression of MLD 6.
    MLD 10.   young adult: biparieto-occipitale (f).   MLD 10a breccia cast of MLD 10.
    MLD 23.   same individual as MLD 6: lt maxilla (ff) with I2, P3.   MLD 23a breccia cast of MLD 23.
    MLD 26.   frontale (ff).   (Probably not hominid.)
    MLD 31.   rt temporale, part of pars mastoidea, posterior half of tympanic plate and exposed middle ear cavity.   (Probably not hominid.)

MLD 33. rt parietale (ff).   (Probably not hominid.)
12.   R. A. Dart 1949, The cranio-facial fragment of *Australopithecus prometheus, Am. J. phys. Anthrop.,* **7** : 187–213 (MLD 3, MLD 6).   E. L. Boné 1955, Un second fragment bipariéto-occipital de l'Australopithèque de Makapansgat (N. Transvaal), *Z. morph. Anthrop.,* **47** : 217–220 (MLD 10).
Undescribed: MLD 31; MLD 26, MLD 33 possibly cercopithecoid.
13.   R. A. Dart 1949 (MLD 3, MLD 6).   E. L. Boné 1955 (MLD 10).   *Australopithecus prometheus.*   J. T. Robinson 1956, The Dentition of the Australopithecinae, *Transv. Mus. Mem.,* **9** : 59, 63, 85, 90.   (MLD 6).
14.   J. T. Robinson 1954a.   *A. africanus transvaalensis.*   J. T. Robinson 1967.   J. T. Robinson 1972.   *Homo africanus.*
15.   E. L. Boné 1956, Un nouveau pariétal droit d'un jeune australopithèque *(A. prometheus)* de Makapansgat, *Z. Morph. Anthrop.,* **48** : 71–78 (MLD 3).   R. A. Dart 1949 (MLD 6).   E. L. Boné 1955 (MLD 10).

1.   Makapansgat Limeworks (1958–1959 cranium).
3.   J. Kitching, December 1958 (MLD 37), April 1959 (MLD 38).
4.   Eroded land surface of Upper Phase I Breccia.
8.   Fauna as in Lower Phase I Breccia, but 80 per cent. baboons.   R. A. Dart 1962, The Makapansgat pink breccia Australopithecine skull, *Am. J. phys. Anthrop.,* **20** : 119–126.
11.   **MLD 37/38.** 'Makapansgat skull' adult female: cranium (f), maxillae (ff) with rt M1 (f), M2, M3 (f), lt M2 (f).
12.   R. A. Dart 1959, A tolerably complete australopithecine cranium from the Makapansgat pink breccia, *S. Afr. J. Sci.,* **55** : 325–327.
13.   R. A. Dart 1962.   Similar to *Plesianthropus* (Sts 5).
14.   J. T. Robinson 1954a.   *A. africanus transvaalensis.*   J. T. Robinson 1967.   J. T. Robinson 1972.   *Homo africanus.*
15.   R. A. Dart 1962.
16.   R. A. Dart 1959, The first *Australopithecus* Cranium from the Pink Breccia at Makapansgat, *Am. J. phys. Anthrop.,* **17** : 77–82.   R. L. Holloway 1973, Endocranial Volumes of Early African Hominids, and the Role of the Brain in Human Mosaic Evolution, *J. hum. Evol.,* **2** : 449–459.
18.   Cast of two blocks containing two parts of cranium, Bernard Price Institute and Department of Anatomy, University of the Witwatersrand, Jan Smuts Avenue, Johannesburg, Transvaal.

1.   Makapansgat Limeworks (1948–1956 maxillae, dentes).
3.   B. Kitching, November 1948 (MLD 9).   A. R. Hughes, July 1949 (MLD 11, MLD 30), 21 August 1956 (MLD 28), J. Kitching, November 1949 (MLD 12, MLD 13).
4.   Limeworks dumps from Lower Phase I Breccia.

11. **MLD 9/12.** adult, female?: rt maxilla (f) with P3, P4, M1, M2, M3 and sockets of I1,
     I2, C, rt palatinum (f).   MLD 9a breccia cast of MLD 9.
     MLD 11/30. juvenile: maxilla (f) in 5 pieces, comprising:
     MLD 11.   (I2), (C), (P3), (P4) (f).
     MLD 30.   lt palatinum and alveolar ridge (f) with M1 (f), (P4) (f) (i.e.
     same tooth as MLD 11, P4).
     MLD 13. isolated, much worn upper C.   (Probably not hominid.)
     MLD 28. rt maxilla (ff) with palatinum (ff) and M2 (f), M3 (probably same in-
     dividual as MLD 4/18).
12. R. A. Dart 1949, A second adult palate of *Australopithecus prometheus, Am. J. phys.
     Anthrop.,* **7** : 335–338 (MLD 9).   J. T. Robinson 1956, The Dentition of the
     Australopithecinae, *Transv. Mus. Mem.,* **9** : 28, 45, 59, 63 (MLD 11).
13. R. A. Dart 1949.   J. T. Robinson 1956: 60, 63, 85, 90 (MLD 9); 28, 45, 59, 63 (MLD
     11). *Australopithecus prometheus.*
14. J. T. Robinson 1954a.   *A. africanus transvaalensis.*   W. E. Le Gros Clark 1964.   *A.
     africanus.*   J. T. Robinson 1967.   J. T. Robinson 1972.   *Homo africanus.*
15. R. A. Dart 1955, The First Australopithecine Fragment from the Makapansgat Pebble
     Culture Stratum, *Nature, Lond.,* **176** : 170, figs 1, 2 (MLD 9).   J. T. Robinson 1956,
     figs 24e, 26f (MLD 9), figs 7b, 7d, 7e, 16g (MLD 11).
18. Department of Anatomy, University of the Witwatersrand, Johannesburg, Medical
     School, Hospital Street, Johannesburg, Transvaal.

1. Makapansgat Limeworks (1948–1961 mandibulae, dentes).
3. A. R. Hughes and S. Kitching, 27 July 1948 (MLD 2); J. Kitching, September 1948
    (MLD 5); A. R. Hughes, 29 July 1953 (MLD 18), 3 August 1955 (MLD 22), 2
    September 1955 (MLD 24), 24 July 1956 (MLD 27); B. Kitching, September 1948
    (MLD 4), 22 September 1956 (MLD 29); E. L. Boné, 21 April 1955 (MLD 19); R. A.
    Dart, 1 September 1960 (MLD 34); B. Maguire, April 1961 (MLD 40).
4. Limeworks dumps from Lower Phase I Breccia.
9. Mandible (MLD 29): F = 0·5%, 100F/P205 = 2·1, eU308 = nil, N = < 0·1%.
11. **MLD 2.** adolescent male, 12 yrs: mandibula, complete from symphysis to lower part
     of lt and rt ramus, with lt P3, (P4), M1, M2, rt I1, I2, (C), (P3), dm2, M1,
     M2.
     MLD 4. lower lt M3 (same individual as MLD 18.   Probably same individual as
     MLD 28).
     MLD 5. juvenile male: lower rt dm2.
     MLD 18. adult female: mandibula (f) with lt I1, I2, C, P3, P4, rt I1, I2, C, P3, P4, M1,
     M2, M3 (same individual as MLD 4.   Probably same individual as MLD
     28).
     MLD 19. adult: mandibula (ff) with M2 distal roots, M3.
     MLD 22. adult, female?: mandibula (ff) with lt M2, M3.   (Lost, cast only.)
     MLD 24. isolated much worn lower lt M2.   MLD 24a breccia cast of MLD
     24.   (Not same individual as MLD 4/18, *pace* R. A. Dart 1962.)
     MLD 27. adult: symphysial region of mandibula (ff) with sockets of lt and rt I1, I2, C,
     roots of lt and rt P3 (possibly same individual as MLD 19).

MLD 29.   adult, male?: mandibula (f) with lt P4, M1 (f).
MLD 34.   mandibula (ff) with rt roots M2, M3 (probably same individual as MLD 22).
MLD 40.   elderly adult: lt mandibula (f) with C (f), P3, P4, M1, M2 and roots of M3.

12.   R. A. Dart 1948, The adolescent mandible of *Australopithecus prometheus, Am. J. phys. Anthrop.,* **6** : 391–411 (MLD 2).   R. A. Dart 1949, The cranio-facial fragment of *Australopithecus prometheus, Am. J. phys. Anthrop.,* **7** : 201–203 (MLD 4, MLD 5).   R. A. Dart 1954, The adult female lower jaw from Makapansgat, *Nature, Lond.,* **173** : 286 (MLD 18).   E. L. Boné 1955, Une clavicule et un nouveau fragment mandibulaire d'*Australopithecus prometheus, Palaeont. afr.,* **3** : 87–101 (MLD 19).   R. A. Dart 1962, A cleft adult mandible and the nine other lower jaw fragments from Makapansgat, *Am. J. phys. Anthrop.,* **20** : 267–286 (MLD 22, MLD 24, MLD 27, MLD 29, MLD 34, MLD 40).

13.   R. A. Dart 1948 (MLD 2).   R. A. Dart 1949 (MLD 4, MLD 5).   *Australopithecus prometheus.*   R. A. Dart 1954, The second, or adult, female mandible of *Australopithecus prometheus, Am. J. phys. Anthrop.,* **12** : 313–343 (MLD 18).   E. L. Boné 1955 (MLD 19).   R. A. Dart 1962 (MLD 22, MLD 27, MLD 29, MLD 34, MLD 40).   J. T. Robinson 1956, The Dentition of the Australopithecinae, *Transv. Mus. Mem.,* **9** : 73, 76, 106, 111, 142 (MLD 2); 116 (MLD 4); 142 (MLD 5); 36, 38, 51, 73, 76, 106, 111, 116 (MLD 18).   P. V. Tobias 1967, *Olduvai Gorge,* Vol. 2, 244, Cambridge.

14.   W. E. Le Gros Clark 1964.   *A. africanus.*   J. T. Robinson 1967.   J. T. Robinson 1972.   *Homo africanus.*   E. Aguirre 1970, Identificación de ' Paranthropus ' en Makapansgat, *XI Congreso Nacional de Arqueologia,* Mérida 1969 : 98–124.   *A. (Paranthropus) robustus* (MLD 2).

15.   R. A. Dart 1948 (MLD 2).   R. A. Dart 1954 (MLD 18).   E. L. Boné 1957, Les Fouilles 1955 au terril de Makapansgat (N. Transvaal), *Proc. 3rd Pan-Afr. Congr. Prehist.* : 149–154, fig 1 (MLD 19).   R. A. Dart 1962, figs 3, 4 (MLD 18, MLD 19, MLD 24, MLD 29, MLD 34), figs 1–10 (MLD 40).   J. T. Robinson 1956, figs 21c, 30d, 32c (MLD 2).

16.   R. A. Dart and E. L. Boné 1955, A Catalogue of the Australopithecine Fossils found at Limeworks, Makapansgat, *Am. J. phys. Anthrop.,* **13** : 623 (MLD 5 wrongly described as lower M2).   M. H. Wolpoff 1973, Posterior tooth size, body size, and diet in South African gracile australopithecines, *Am. J. phys. Anthrop.,* **39** : 375–393.

1.   Makapansgat Limeworks (1948–1960 post-cranial bones).

3.   J. Kitching, November 1948 (MLD 7, MLD 8), 30 April 1956 (MLD 32), December 1958 (MLD 39); A. R. Hughes, November 1949 (MLD 14, MLD 15, MLD 16, MLD 17), 27 April 1955 (MLD 20), 3 May 1955 (MLD 21); R. A. Dart, 1 September 1960 MLD 36).

4.   Limeworks dumps from Lower Phase I Breccia.

7.   R. A. Dart 1961, An australopithecine scoop made from a right australopithecine upper arm bone, *Nature, Lond.,* **191** : 372–373.

11.   **MLD 7.**   adolescent male: lt ilium (possibly same individual as MLD 2).
       MLD 8.   rt ischial fragment (probably same individual as MLD 7).

> **MLD 14.** rt humerus (f), distal portion of diaphysis.
> **MLD 15.** rt radius (f), distal portion of diaphysis.   (Probably not hominid.)
> **MLD 16.** rt radius (f), proximal extremity.   (Probably not hominid.)
> **MLD 17.** head of femur (f).   (Probably not hominid.)
> **MLD 20.** rt clavicula (f), acromial extremity.   MLD 20a breccia cast of MLD 20.   (Not hominid.)
> **MLD 21.** ilium ? (ff).   MLD 21a breccia cast of MLD 21.   Bone indet. (? Hominid.)
> **MLD 25.** adolescent female: lt ilium (f).
> **MLD 32.** lt radius (f), proximal extremity.   (Probably not hominid.)
> **MLD 35.** infant: ilium ? (ff).   (Probably not hominid.)
> **MLD 36.** lt clavicula (f), acromial extremity.   (Not hominid.)
> **MLD 39.** rt humerus (f), distal portion, spirally broken.   (Not hominid.)

12. R. A. Dart 1949, The first pelvic bones of *Australopithecus prometheus:* preliminary note, *Am. J. phys. Anthrop.,* **7** : 255–258.   R. A. Dart 1958, A further adolescent australopithecine ilium from Makapansgat, *Am. J. phys. Anthrop.,* **16** : 473–479 (MLD 25).   E. L. Boné 1955, Quatre fragments post-craniens du gisement à Australopithèques de Makapansgat (N. Transvaal), *Anthropologie, Paris,* **59** : 462–469 (MLD 14, MLD 15, MLD 16, MLD 17).   E. L. Boné 1955, Une clavicule et un nouveau fragment mandibulaire d'*Australopithecus prometheus, Palaeont. afr.,* **3** : 87–101 (MLD 20).   R. A. Dart 1961 (MLD 39).

13. R. A. Dart 1949, Innominate fragments of *Australopithecus prometheus, Am. J. phys. Anthrop.,* **7** : 301–334 (MLD 7, MLD 8).   R. A. Dart 1957, The second adolescent (female) ilium of *Australopithecus prometheus, J. palaeont. Soc. India,* **2** : 73–82 (MLD 25).   Undescribed: MLD 32, MLD 35, MLD 36.

14. J. T. Robinson 1972, *Early Hominid Posture and Locomotion:* 1–361, Chicago and London (MLD 7, MLD 8, MLD 14, MLD 20).   D. A. Hooijer 1975, Miocene to Pleistocene Hipparions of Kenya, Tanzania and Ethiopia, *Zool. Verh. Leiden,* No. 142 : 50–51, Pl 19 fig. 3, 4 (MLD 20, MLD 36).   *Hipparion steytleri.*

15. R. A. Dart 1949 (MLD 7).   R. A. Dart 1958 (MLD 7, MLD 25).   R. A. Dart 1961: 372–373 (MLD 14).

16. W. E. Le Gros Clark 1955, The os innominatum of the recent Ponginae with special reference to that of the Australopithecinae, *Am. J. phys. Anthrop.,* **13** : 19–27.   J. R. Napier 1964, The Evolution of Bipedal Walking in the Hominids, *Arch. Biol.* (Liège), **75** suppl. : 673–708 (MLD 7).

18. Department of Anatomy, University of the Witwatersrand, Johannesburg, Medical School, Hospital Street, Johannesburg, Transvaal.

MATJES RIVER

1. Matjes River.
2. Rock-shelter, 60 m above Matjes River, about 13 km NE of Plettenburg Bay, Cape Province.   34° 00′ S, 23° 27′ E.
3. T. F. Dreyer, 1930–32.
4. Sterile layer in occupation deposits (layer E).   A. Keith 1933, A Descriptive Account of the Human Remains from Matjes River Cave, Cape Province, *Trans. R. Soc. S. Afr.,*

21 : 151–185.   J. T. Louw 1960, Prehistory of the Matjes River Rock-shelter, *Nas. Mus. Bloemfontein Mem.,* **1** : 1–143.

5.   Burials from layer D.   J. T. Louw 1960.

6.   Late Pleistocene/Holocene.

7.   Beginning of Later Stone Age (Smithfield).   R. J. Mason 1962, *Prehistory of the Transvaal,* Johannesburg.   J. T. Louw 1960.

8.   *Phacochoerus 'dreyeri'* (species invalid, H. B. S. Cooke, *in lit.)* and modern species.   T. F. Dreyer and A. Lyle 1931, *New Fossil Mammals and Man from South Africa,* Bloemfontein.

9.   ————

10.   A2: 10,500 ± 400 BP (L-336G); 11,250 ± 400 BP (L-336H) on basis of C14 dating of ? charcoal from layer D.   J. T. Louw 1960.   Shelter was occupied from before 8100 BC to at least 1600 BC, on the basis of C14 dating of stratigraphic series of shell and charcoal samples from the rock shelter.   J. C. Vogel 1970, *Radiocarbon,* **12** : 463.   R. Protsch and J. J. Oberholzer 1975, Palaeoanthropology, chronology and archaeology of the Matjes River Rock Shelter, *Z. Morph. Anthrop.,* **67** : 32–43.

11.   18 individuals of both sexes and varying ages.   Crania and mandibulae well preserved, and some post-cranial bones.

12.   T. F. Dreyer and A. Lyle 1931.

13.   A. Keith 1933.

14.   L. H. Wells 1952, Human Crania in the Middle Stone Age in South Africa, *Proc. 1st Pan-Afr. Congr. Prehist.* : 125–133.

15.   A. Keith 1933.

16.   A. J. D. Meiring 1937, The Frontal Convolutions on the Endocranial Cast of the Skull M.R.I. from the deepest levels of the Matjes River cave, *S. Afr. J. Sci.,* **33** : 960–970.   T. F. Dreyer, A. C. Hoffman and A. J. D. Meiring 1938, A comparison of the Boskop with other abnormal skull-forms from South Africa, *Z. Rassenk.,* **7** : 289–296.   L. H. Wells 1959, The Problem of Middle Stone Age Man in Southern Africa, *Man,* **59** : 158–160.

17.   National Museum, Bloemfontein, Orange Free State.

18.   ————

NAHOON POINT   33° 00′ S, 27° 52′ E.

In calcareous sandstone of the cliffs, about 5 km E of East London, Cape Province, R. Kaiser and W. Hartley discovered three 'footprints' (casts) of *Homo* in July 1964.   Footprints of hyaena and birds were associated.   Probable correlation with artifacts of Middle Stone Age in Bat's Cave 2 km to W.   Molluscan shell in the sandstone gave a C14 age of 29,090 + 410 − 330 BP (SR–83).

E. D. Mountain 1966, Footprints in Calcareous Sandstone at Nahoon Point, *S. Afr. J. Sci.,* **62** : 103–111.   H. J. Deacon 1966, The Dating of the Nahoon Footprints, *S. Afr. J. Sci.,* **62** : 111–113.

The cast is preserved in the East London Museum, Upper Oxford St, East London, Cape Province.

NELSON BAY   34° 06′ S, 23° 24′ E.

In 1970, R. G. Klein discovered human remains in shell middens in a wave-cut cave at Nelson Bay on Robberg Peninsula, Plettenburg Bay, Cape Province.   The middens contained Late Stone Age industries and extensive faunal remains of mammals, birds and fishes.   The data are listed in the following Table.

R. G. Klein 1972a, Preliminary Report on the June–September 1970 excavations at Nelson Bay Cave (Plettenburg Bay, Cape), *Palaeoecology of Africa,* Vol. 6, Cape Town : 177–208.   R. G. Klein 1972b, The Late Quaternary Mammalian Fauna of Nelson Bay Cave (Cape Province, South Africa); its Implications for Megafaunal Extinctions and Environmental and Cultural Change, *Quatern. Res.,* 2 (2) : 135–142.   R. G. Klein 1974, A provisional statement of Terminal Pleistocene Mammalian Extinctions in the Cape biotic zone (Southern Cape Province, South Africa), *S. Afr. archaeol. Soc., Goodwin Ser.,* 2 : 32–45.

C14 DATE	MIDDEN	STRATIGRAPHICAL UNIT	INDUSTRY	HOMINIDS
A2: 5350 ± 65 BP (UW-217) on *Patella* shell	Ivan (NB4, NB5) Betsy (NB6)	IC	Wilton	NB4: lt calcaneus NB5: rt calcaneus Associated fragments: mandibula (f), 11 dentes, M (germ), proximal lt femur (ff), distal rt tibia (f), vertebra thoracicalis. NB6: prox. ulna
A2: 5825 ± 150 BP (UW-187) on charcoal	Brown soil below Middens ‘ Glen/Helgren ’	BSC	Wilton	NB7: M.
A2: 8120 ± 240 BP (UW-181) 8990 ± 80 BP (Pta-391) on charcoal	Jake	J	Wilton	NB9: M (crown)
A2: 8570 ± 170 BP (UW-184) on *Patella* shell	Rice B (NB8, NB12 burial) Chris (NB13)	RC	Wilton	NB8: maxilla (ff), P4, M1, M2. NB12: cranium (f), skeleton (i) (left *in situ*). NB13: 2M, phalanx (possibly same individual as NB12).

OAKHURST

1. Oakhurst.
2. 21 km E of George, Cape Province.   34° 00′ S, 22° 45′ E.
3. A. J. H. Goodwin, 1932–1935.

4.   Cave occupation deposits.   A. J. H. Goodwin 1937, The Archaeology of the Oakhurst Shelter, George, *Trans. R. Soc. S. Afr.,* **25** : 303–320.
5.   Succession of burials, many of the earlier disturbed by the later.   A. J. H. Goodwin 1937 : 247–258.
6.   Holocene.
7.   Later Stone Age, Wilton.   A. J. H. Goodwin 1937 : 229–258, 303–324.   J. F. Schofield 1937, The Archaeology of the Oakhurst Shelter, *Trans. R. Soc. S. Afr.,* **25** : 295–302.
8.   Recent fauna.
9.   ———
10.   A3: *c.* 6000 BP on basis of C14 datings of Wilton carbon from layer C, Matjes River: 7750 ± 350 BP (L–336E) and 5400 ± 250 BP (L–336F).
11.   Remains of 12 adults: 8 male, 4 female and 15 infants and children under 7 years old.
12.   A. J. H. Goodwin 1937.
13.   M. R. Drennan 1937, The Archaeology of the Oakhurst Shelter, George.   III The Cave-Dwellers: IV The Children of the Cave Dwellers, *Trans. R. Soc. S. Afr.,* **25** : 259–294.   Oakhurst tribe.
14.   ———
15.   A. J. H. Goodwin 1937.
16.   ———
17.   Department of Anatomy, University of Cape Town, P.O. Box 594, Rondebosch. Cape Province.
18.   ———

PEERS' CAVE

See under **FISH HOEK**

PHILIPPI

See under **CAPE FLATS**

SALDANHA

See under **HOPEFIELD**

SKILDERGAT

See under **FISH HOEK**

SPRINGBOK FLATS

See under **TUINPLAAS**

STERKFONTEIN

General Entry

1. Sterkfontein.
2. Exposed breccia filling of ancient cave in dolomitic limestone, Sterkfontein valley, 13 km NNW of Krugersdorp, Transvaal.   26° 03′ S, 27° 42′ E.
3. Members of the Transvaal Museum, Pretoria, 1936–1958.   Members of the Department of Anatomy, University of the Witwatersrand, Johannesburg, 1966–1973.
4. Cave breccia exposed and excavated in different places, following travertine mining (1930–1945) (Fig. 2).   Hominid fossils are derived from Member IV (Lower Breccia) and Member V (Middle Breccia).   Most of the exposures in the main quarry (formerly called 'Type Site') and in Robert Broom's 'Lower Cave' have been assigned to Member 4 of the Sterkfontein Formation, as have the materials in Dumps 9, 10, 13, 14, 15, 17 and 18, probably Dump 12 and the easterly $\frac{1}{3}$ of Dump 8: these dumps resulted from earlier activities by lime workers and by palaeontologists.   Most of the exposures in the West Pit (formerly called 'Extension Site') as well as in recently exposed parts south and west of the West Pit have been assigned to Member 5 of the Sterkfontein Formation, as have the materials in Dumps 1 and 3.   C. K. Brain 1958, The Transvaal Ape-Man-Bearing Cave Deposits, *Transv. Mus. Mem.,* **11** : 72–73.   J. T. Robinson

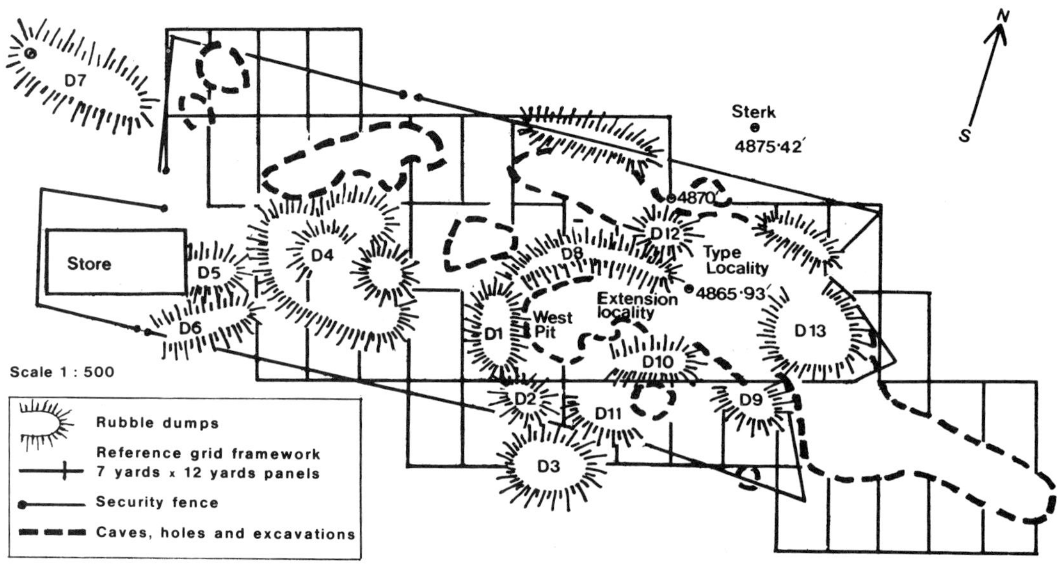

Fig. 2.   Plan of the Sterkfontein excavation site. (From *S. Afr. archaeol. Bull.,* **24** : 164.)

1962, Australopithecines and Artifacts at Sterkfontein, pt I, Sterkfontein Stratigraphy and the Significance of the Extension site, *S. Afr. archaeol. Bull.,* **17** : 97.   P. V. Tobias and A. R. Hughes 1969, The New Witwatersrand University Excavation at Sterkfontein, *S. Afr. archaeol. Bull.,* **24** : 158–169.   K. W. Butzer 1971, Another look at the australopithecine cave breccias of the Transvaal, *Am. Anthrop.,* **73** : 1197–1201.

5. ————

6. Late Villafranchian.   K. P. Oakley 1957, Dating the Australopithecines, *Proc. 3rd Pan-Afr. Congr. Prehist.* : 155–157.
First Interpluvial.   C. K. Brain 1958, The Transvaal Ape-Man-Bearing Cave Deposits, *Transv. Mus. Mem.,* **11** : 73.
Sterkfontein Faunal Span.   H. B. S. Cooke 1968, Evolution of Mammals on Southern Continents. II. The Fossil Mammals of Africa. *Quart. Rev. Biol.,* **43** (3) : 234–264.   L. H. Wells 1969, Faunal subdivision of the Quaternary in southern Africa, *S. Afr. Archaeol. Bull.,* **24** : 93–95.

7. Industry reported only from the second highest breccia of the West pit, Member 5, formerly ' Extension ' locality.   Developed Oldowan or earliest stage of African-Acheulian.   J. T. Robinson and R. J. Mason 1957, Occurrence of Stone Artefacts with *Australopithecus* at Sterkfontein, *Nature, Lond.,* **180** : 521–524.   C. K. Brain 1958.   R. Mason 1962, Australopithecines and Artefacts at Sterkfontein, pt. 2, The Sterkfontein Stone Artefacts and their Maker, *S. Afr. archaeol. Bull.,* **17** : 109–125.   M. D. Leakey 1970, Stone artefacts from Swartkrans, *Nature, Lond.,* **225** : 1222–1225.

8. Member 4: Sterkfontein Faunal Span, including *Parapapio broomi, P. whitei, P. jonesi, Cercopithecoides williamsi, Dinofelis barlowi, Canis brevirostris, Lycyaena, Procavia transvaalensis, P. antiqua, Equus helmei, Makapania broomi, Redunca darti, Tapinochoerus.* R. F. Ewer 1957, Faunal evidence on the dating of the Australopithecinae, *Proc. 3rd Pan-Afr. Congr. Prehist.* : 135–142.   L. H. Wells 1962, Pleistocene faunas and the distribution of mammals in Southern Africa, *Ann. Cape Prov. Mus.,* **2** : 37–40.   L. H. Wells 1969.   H. B. S. Cooke 1963, 1968.   E. S. Vrba 1974, Chronological and Ecological implications of the fossil Bovidae at the Sterkfontein australopithecine site, *Nature, Lond.,* **250** : 19–23.   Member 5: Cornelia Faunal Span.   E. S. Vrba 1975, Some evidence of chronology and palaeoecology of Sterkfontein, Swartkrans and Kromdraai from the fossil Bovidae, *Nature, Lond.,* **254** : 301–304.   E. S. Vrba 1976, The Fossil Bovidae of Sterkfontein, Swartkrans and Kromdraai, *Transv. Mus. Mem.,* **21** : 1–166.

9. ————

10. T. C. Partridge 1973, Geomorphological Dating of Cave Openings at Makapansgat, Sterkfontein, Swartkrans and Taung, *Nature, Lond.,* **246** : 75–79.

11.–15.   See under individual entries.

16. J. T. Robinson 1954, The Genera and Species of the Australopithecinae, *Am. J. phys. Anthrop.,* **12** : 196. *Australopithecus africanus transvaalensis.* W. E. Le Gros Clark 1964, *The Fossil Evidence for Human Evolution,* 2nd edn Chicago : 123–173. *Australopithecus africanus.* W. E. Le Gros Clark, 1967, *Ape-Man or Man-Ape?,* New York and London.   H. B. S. Cooke 1969, Preservation of the Sterkfontein Ape-man Cave Site, South Africa, *Stud. Speleol.,* **2** (1), 25–34.   P. V. Tobias 1972, Progress and problems in the study of early man in sub-Saharan Africa, *in* R. Tuttle (ed.) 1972, *The Functional and Evolutionary Biology of Primates,* Chicago : 63–93.   J. T. Robinson 1972, *Early Hominid Posture and Locomotion:*

1–361 Chicago and London.   P. V. Tobias 1973, Implications of the New Age Estimates of the Early South African Hominids, *Nature, Lond.,* **246** : 79–83.   R. L. Holloway 1973, Endocranial Volumes of Early African Hominids, and the Role of the Brain in Human Mosaic Evolution, *J. hum. Evol.,* **2** : 449–459 (Sts 60, 71, 19/58, 5).   H. M. McHenry 1974, How large were the Australopithecines?, *Am. J. phys. Anthrop.,* **40** : 329–340 (Sts 7, 14, 34, 68, TM 1513).   A. E. Mann 1975, Some paleodemographic aspects of the South African australopithecines, *Univ. Pennsylv. Publs Anthrop.,* **1** : 1–171.

17.   Transvaal Museum, P.O. Box 413, Pretoria, Transvaal (Sts & TM numbers), Department of Anatomy, University of the Witwatersrand, Johannesburg, Medical School, Hospital Street, Johannesburg (StW numbers).

18.   While stocks last, casts may be bought from the Wenner-Gren Foundation, 14 East 71st Street, New York, N.Y. 10021, U.S.A.   (Sts 5, Sts 7, Sts 8, Sts 14, Sts 24, Sts 34, Sts 52b, Sts 60, Sts 68, TM 1512, TM 1513.)

Index of Fossil Hominid Remains from Sterkfontein

Reg. No.	Element	Page	Reg. No.	Element	Page
TM 1511	cranium	122	Sts 9	M1	129
TM 1512	maxilla	124	Sts 10	maxilla	124
TM 1513	femur	131	Sts 11	M1, M2	127
TM 1514	maxilla	124	Sts 12	maxilla	124
TM 1515	mandibula	126	Sts 13	ossa faciei	124
TM 1516	mandibula	126	Sts 14	vertebrae, os coxae, femur	131
TM 1518	M3	129			
TM 1519	M3	129	Sts 16	M	127
TM 1520	M3	129	Sts 17	cranium	123
TM 1522	mandibula	126	Sts 18	mandibula	126
TM 1523	P4	129	Sts 19	basis cranii	123
TM 1524	M1	127	Sts 20	calvaria	123
TM 1526	os capitatum	131	Sts 21	M1	127
TM 1527	C	127	Sts 22	M1, M2	127
TM 1528	C	129	Sts 23	M2	127
TM 1532	M3	127	Sts 24	mandibula	126
TM 1534	I	129	Sts 25	calvaria	123
TM 1535	C	124	Sts 26	cranium	123
TM 1561	M3	127	Sts 27	maxilla	124
			Sts 28	M1, M2, M3	127
Sts 1	palatinum	125	Sts 29	palatinum	125
Sts 2	ossa faciei, maxilla	124	Sts 30	P4, M2	127
Sts 3	C	129	Sts 31	M3	127
Sts 4	M2	129	Sts 32	maxilla	124
Sts 5	cranium	122	Sts 33	P	127
Sts 6	M1, M2	122	Sts 34	femur	131
Sts 7	ramus mandibulae, scapula	126	Sts 35	maxilla	124
			Sts 36	mandibula	126
Sts 8	maxilla	124	Sts 37	M2, M3	127

Reg. No.	Element	Page
Sts 38	mandibula	126
Sts 39	P4, M1	127
Sts 40	C	129
Sts 41	mandibula	126
Sts 42	palatinum	125
Sts 43	M1	127
Sts 44	M2, M3	127
Sts 45	M	127
Sts 46	M3	127
Sts 47	P3	127
Sts 48	C	127
Sts 49	P3	129
Sts 50	C	126
Sts 51	P3, C	129
Sts 52a	ossa faciei	124
Sts 52b	mandibula	124
Sts 53	palatinum	125
Sts 54	M3	127
Sts 55	P3	127
Sts 55b	M3	129
Sts 56	M1, M2, dm2	127
Sts 57	maxilla	124
Sts 58	calvaria	123
Sts 59	M3	129
Sts 60	endocranial cast	124
Sts 61	maxilla	124
Sts 62	mandibula	126
Sts 63	ossa faciei	124
Sts 64	palatinum	125
Sts 65	vertebrae, os coxae	131
Sts 66	palatinum	124
Sts 67	calvaria	123
Sts 68	radius	131
Sts 69	maxilla	124
Sts 70	maxilla	124
Sts 71	cranium	123
Sts 72	M3	127
Sts 73	vertebrae	131
Sts 3009	maxilla	124
StW/H 1	M1	129
StW/H 2	M3	129
StW/H 3	M3	129
StW/H 4	M1/M2	129
StW/H 5	I	129
StW/H 6	M3	129
StW/H 7	P3	129
StW/H 8	vertebrae	131
StW/H 9A	P4	129
StW/H 9B	M1	129
StW/H 10	P	129

Reg. No.	Element	Page
StW/H 11	M	129
StW/H 12	M3	123
StW/H 13	calvaria	123
StW/H 14	mandibula	127
StW/H 15	I1	128
StW/H 16	P	129
StW/H 17	maxilla	123
StW/H 18	maxilla	129
StW/H19a	P	133
StW/H19b	M2, M3	133
StW/H20	C	130
StW/H21	C	130
StW/H22	C (root)	130
StW/H23	I2	130
StW/H24	I2	130
StW/H25	femur	132
StW/H26	metacarpale	132
StW/H27	metacarpale	133
StW/H28	phalanx manus	132
StW/H29	phalanx manus	132
StW/H30a	femur	132
StW/H30b	femur	132
StW/H30c,d,e	os coxae	132
StW/H31	femur	132
StW/H32	I 1	130
StW/H35	P	130
StW/H36	maxilla	130
StW/H37	M3	130
StW/H38	humerus	132
StW/H39	mandibula	127
StW/H40	maxilla	125
StW/H41	vertebrae	131
StW/H42	I2	133
StW/H43	M3	130
StW/H44	C	130
StW/H45	P	130
StW/H46	radius	132
StW/H47	M3	130
StW/H48	M2	130
StW/H49	maxilla	125
StW/H50	P3	130
StW/H51	M2	130
StW/H52	M2 or M3	130
StW/H53	cranium, mandibular ramus	133
StW/H54	M	130
StW/H55	P3	130
StW/H56	P4	130
StW/H57	P4	130
StW/H58	C	130
StW/H59	maxilla	125

Reg. No.	*Element*	*Page*	*Reg. No.*	*Element*	*Page*
StW/H60	P	130	SE 2396	P3	132
StW/H61	M	130	SE 255	maxilla	132
StW/H62	di	130	SE 1508	M2	132
StW/H63	metacarpale	132	SE 1579	M1/M2	132
StW/H64	metacarpale	132	SE 1937	C	132

MEMBER 4

1. Sterkfontein, ' Type Site ' (1936 cranium).
3. R. Broom, 17 August 1936.
11. **TM 1511.** young adult male: cranium (f) including nasalia (f), occipitale (ff), maxillae (f) with rt P4, M1, M2 and lt P3, P4, M1, M2; isolated rt M3.   **Holotype** of *Plesianthropus transvaalensis* Broom, 1936.   Plate 10.
     Also natural endocranial cast (f).
12. R. Broom 1936a, A new fossil anthropoid skull from South Africa, *Nature, Lond.,* **138** : 486–488.   R. Broom 1936b, The dentition of *Australopithecus, Nature, Lond.,* **138** : 719.
13. R. Broom and G. W. H. Schepers 1946, The South African Fossil Ape-Men: the Australopithecinae, *Transv. Mus. Mem.,* **2** : 58, 62, 84, 88, 93. *Plesianthropus transvaalensis.*
14. ————
15. R. Broom and G. W. H. Schepers 1946.   W. K. Gregory and M. Hellman 1945, Revised reconstruction of the skull of *Plesianthropus transvaalensis* Broom, *Am. J. phys. Anthrop.,* **3** : 267–275.   J. T. Robinson 1956, The Dentition of the Australopithecinae, *Transv. Mus. Mem.,* **9** ; fig. 28e.

1. Sterkfontein, ' Type Site ' (1947 cranium).
3. R. Broom and J. T. Robinson, 18 April 1947.
11. **Sts 5.** ' Mrs Ples ', adult: virtually complete and undistorted cranium.   No dentes (possibly same individual as Sts 6 or Sts 7).
     Sts 6. lower M2 and M1 (f) (possibly belonging to Sts 5.)
12. R. Broom 1947, Discovery of a new skull of the South African ape-man, *Plesianthropus, Nature, Lond.,* **159** : 672.
13. R. Broom, J. T. Robinson and G. W. H. Schepers 1950, Sterkfontein Ape-Man *Plesianthropus, Transv. Mus. Mem.,* **4** : 14–24.   J. T. Robinson 1956, The Dentition of the Australopithecinae, *Transv. Mus. Mem.,* **9** : 110 (Sts 6).   *Plesianthropus transvaalensis.*

14. ————

15. R. Broom, *et al.* 1950.   pl. 1, figs 1 and 2, pl. 2, figs 6 and 7, pl. 4, fig. 16.

1. Sterkfontein, ' Type Site ' (1947–1949).
3. R. Broom and J. T. Robinson, 26 September 1947 (Sts 17), 31 October 1947 (Sts 19), 4 November 1947 (Sts 20), 23 March 1948 (Sts 25), 2 April 1948 (Sts 26), 31 October 1947 (Sts 58), 13 November 1947 (Sts 71); J. T. Robinson, 1949 (Sts 67).
11. **Sts 17.**   adult female: cranium (f) including part of parietale, lt ossa faciei, palatinum with lt P3, P4, M1 (f), rt P4, M1, M2, M3 (f).
    Sts 19.   adult: cranial base with foramen magnum (i), with lt M3 (f) (possibly same individual as Sts 58).
    Sts 20.   calvaria (ff).
    Sts 25.   adult: calvaria including frontale (ff), parietalia, occipitale (f), temporalia (ff, f).
    Sts 26.   adult, probably female: (two pieces) cranium (ff) including part of frontale, parietale, occipitale with foramen magnum, temporale, lt maxilla (ff) with M3.
    Sts 58.   adult: calvaria (ff) (possibly same individual as Sts 19).
    Sts 67.   calvaria (ff) with region of external auditory meatus; associated upper lt M3.
    Sts 71.   adult: rt side of cranium (f), P3 (ff), P4 (f), M1 (f) M2, (f), M3 (f).
12. R. Broom, J. T. Robinson and G. W. H. Schepers 1950, Sterkfontein Ape-Man *Plesianthropus, Transv. Mus. Mem.,* **4.**
13. R. Broom, *et al.* 1950 : 24–25 (Sts 17 ' skull 6 '); : 27–33 (Sts 19 and Sts 58 ' skull 8 '); 26 (Sts 71 ' skull 7 ').   Transvaal Museum Catalogue (unpublished) (Sts 20, Sts 25, Sts 26, Sts 67). *Plesianthropus transvaalensis.*
14. ————
15. R. Broom, *et al.* 1950 (Sts 17, Sts 19, Sts 58, Sts 71).

1. Sterkfontein, ' Type site ' (1971 cranium).
3. Stephaans Gasela, A. R. Hughes and P. V. Tobias, 8 October 1971 (StW/H 13); A. R. Hughes and P. V. Tobias, 27 August 1971 (StW/H 12); A. R. Hughes and P. V. Tobias, 12 April 1972 (StW/H 17).
4. In breccia E of type locality and SE of the East Pit, 9 m ESE of position from which Sts 5 was recovered.   P. V. Tobias 1972, Summary Progress Report on Palaeo-Anthropological Programme during 1971 (Departmental Report).
11. **StW/H 12.**   upper lt M3, worn (f) (part of the M3 in maxilla of StW/H 13).
    StW/H 13.   adult: calvaria (f), basis cranii (f); almost complete face and palate, rt I2, P3 (ff); lt I2 (ff), P3, P4, M1, M2, M3 (f).
    StW/H 17.   rt maxilla (f) with buccal halves of rt P4, M1 worn (belongs with StW/H 13).
12. P. V. Tobias 1973, Darwin's Prediction and the African Emergence of the Genus *Homo, in* L'origine dell'Uomo, *Accademia Nazionale dei Lincei,* **182** : 63–85, Rome.
13. ————

14. ———————
15. P. V. Tobias 1973.
16. P. V. Tobias 1974, A new chapter in the history of the Sterkfontein early hominid site, *J. S. Afr. biol. Soc.,* **14** : 30–44.

1. Sterkfontein ' Type Site ' (1947, 1949 ossa faciei).
3. R. Broom and J. T. Robinson, 3 and 8 April 1947 (Sts 2), 29 July 1947 (Sts 13); J. T. Robinson, 1949 (Sts 52, Sts 63).
11. **Sts 2.** juvenile: ossa faciei with dm1, dm2, lt (c); maxilla (f) with rt dm1, dm2 (f), (I).
    Sts 13. elderly adult: ossa faciei with most of rt dentition and part of lt dentition. (Specimen missing).
    Sts 52. adolescent, probably female: Sts 52a ossa faciei with maxillae and full upper dentition; Sts 52b mandibula with full dentition but lacking lt ascending ramus.
    Sts 63. elderly adult: rt ossa faciei (ff) with orbit and some fragmentary dentes.
12. R. Broom, J. T. Robinson and G. W. H. Schepers 1950, Sterkfontein Ape-Man *Plesianthropus, Transv. Mus. Mem.,* **4** (Sts 2). R. A. Dart 1954, The second, or adult, female mandible of *Australopithecus prometheus, Am. J. phys. Anthrop.,* **12** : 326 (Sts 52). Transvaal Museum Catalogue (unpublished) (Sts 13, Sts 63).
13. R. Broom, J. T. Robinson and G. W. H. Schepers 1950: 48 (Sts 2). *Plesianthropus transvaalensis.* J. T. Robinson 1956, The Dentition of the Australopithecinae, *Transv. Mus. Mem.,* **9** : 44, 124, 128, (Sts 2); 24, 28, 44, 58, 62, 84, 88, 93, (Sts 52a); 28, 36, 48, 71, 75, 104, 107, 114 (Sts 52b).
14. ———————
15. R. Broom *et al.* 1950 (Sts 2, Sts 52a, Sts 52b). R. A. Dart 1962, A Cleft Adult Mandible and the Nine other Lower Jaw Fragments from Makapansgat, *Am. J. phys. Anthrop.,* **20** : 267–286 (Sts 52b).

1. Sterkfontein ' Type Site ' (1936–1949 maxillae).
3. R. Broom, 1936 (TM 1512 and TM 1514), 1938 (TM 1535), R. Broom and J. T. Robinson, 24 June 1947 (Sts 8), 4 July 1947 (Sts 10), 18 July 1947 (Sts 12), 11 May 1948 (Sts 27), 28 May 1948 (Sts 32), 13 July 1948 (Sts 35); J. T. Robinson, 1949 (Sts 57, Sts 61, Sts 66, Sts 69, Sts 70); C. K. Brain, developed from 1947–1949 breccia in laboratory, in 1968 (Sts 3009).
11. **TM 1512.** adult, probably female: rt maxilla with I2, C, P3, M1.
    TM 1514. old adult, probably male: lt maxilla (f) with worn and broken teeth, M3 reasonably preserved.
    TM 1535. upper C of TM 1514.
    Sts 8. lt maxilla (ff) with M1, M2, M3.
    Sts 10. maxilla (ff) with M2 (f), M3.
    Sts 12. maxilla (f) with lt P3, P4, M1, M2, M3 (f); rt P4, M1, M2.
    Sts 27. maxilla (ff) with M2, M3. Specimen missing.

Sts 32.　maxilla (ff) with M2, M3 (f).
Sts 35.　maxilla (f) with lt P3, P4, M1, M2.
Sts 57.　maxilla (ff) with lt M1 and unerupted crowns of P3, P4.
Sts 61.　maxilla (ff) with P3, P4, M1, M2 (f)
Sts 66.　palatinum, maxilla (ff) with traces of I1–M3
Sts 69.　juvenile; maxilla (ff) with 2I1, I2, and fragments.
Sts 70.　juvenile: maxilla (f) with some deciduous dentes and some unerupted anterior permanent dentes.
Sts 3009.　old adult: rt maxilla (ff) with rt P4 (roots only), M1 (f), M2 (f) and M3, and many cranial and dental fragments.

12.　W. K. Gregory and M. Hellman 1939, The dentition of the extinct South African Man-Ape *Australopithecus (Plesianthropus) transvaalensis* Broom, *Ann. Transv. Mus.,* **19** : 339–373 (TM 1512).　R. Broom and G. W. H. Schepers 1946, The South African Fossil Ape-Men: the Australopithecinae, *Transv. Mus. Mem.,* **2** : 49, 55, 57–60 (TM 1512) 49–50, 55, 56 (TM 1514) 56 (TM 1535).　Transvaal Museum Catalogue (unpublished) (Sts 8, Sts 10, Sts 27, Sts 35, Sts 61, Sts 66, Sts 69, Sts 70, Sts 3009).

13.　R. Broom and G. W. H. Schepers 1946 : 49, 55, 57–60 (TM 1512) 49–50, 55, 56 (TM 1514) 56 (TM 1535). *Plesianthropus transvaalensis.*　J. T. Robinson 1956, The Dentition of the Australopithecinae, *Transv. Mus. Mem.,* **9** : 28, 44, 58, 84 (TM 1512) 48, 88 (Sts 8) 57, 88 (Sts 12) 88 (Sts 32) 84 (Sts 57).

14.　———————

15.　R. Broom and G. W. H. Schepers 1946.　W. K. Gregory and M. Hellman 1939.　J. T. Robinson, 1956.

1.　Sterkfontein Lower Breccia, ' Type Site ' (1947–1949 palatina).
3.　R. Broom and J. T. Robinson, 3 April 1947 (Sts 1), 16 June 1948 (Sts 29), 1948 (Sts 42).　J. T. Robinson, 1949 (Sts 53, Sts 64).
11.　**Sts　1.**　adult: crushed palatinum with lt P3, M1, M2 (f); rt P3, M1 (f).
Sts 29.　adult: palatinum (ff) with rt M (f), M2, M3 (f).
Sts 42.　adult: palatinum (f) with rt P3–M3 and badly crushed pieces of lt maxilla.
Sts 53.　adult: palatinum with rt and lt P3 roots, P4 (f), M1–M3 (i).
Sts 64.　elderly adult: palatinum (ff) with some dentes (ff).

12.　J. T. Robinson 1956, The Dentition of the Australopithecinae, *Transv. Mus. Mem.,* **9** : 58, 84, 88 (Sts 1, Sts 42, Sts 53).　Transvaal Museum Catalogue (unpublished) (Sts 29, Sts 64).

13.　J. T. Robinson 1956: 58, 84 (Sts 1): 58, 62 (Sts 42): 88, 93 (Sts 53).

14.　———————

15.　J. T. Robinson 1956, figs 16a, 16f, 24a, 26e (Sts 1), fig. 28d (Sts 53, wrongly given as Sts 33).

1.　Sterkfontein, dumps derived from Member 4 (1973–1976 maxillae).
11.　**StW/H 40.**
StW/H 49.
StW/H 59.

  1.   Sterkfontein, ' Type Site ' (1936–1938 mandibulae).

  3.   R. Broom, 1938.

11.   **TM 1515.**   elderly adult, probably male: mandibula (ff) with lt P3 (f), P4 (f), M1 (f), M2.

        TM 1516.   juvenile: symphysial region of mandibula (ff) with lt (C) and rt (C) (f).   Lower dm1 (f) separate.

        Sts 50.   lower lt (C) from TM 1516.

        TM 1522.   rt ramus mandibulae (f), M3 (roots).

12.   W. K. Gregory and M. Hellman 1939, The dentition of the extinct South African Man-Ape *Australopithecus (Plesianthropus) transvaalensis*, Broom, *Ann. Transv. Mus.*, **19** : 346 (TM 1516, Sts 50).   Transvaal Museum Catalogue (unpublished) (TM 1522).

13.   R. Broom and G. W. H. Schepers, 1946, The South African Fossil Ape-Men: the Australopithecinae, *Transv. Mus. Mem.*, **2** : 54, 67, 69 (TM 1515): 53, 65, 72 (TM 1516): 65 (Sts 50).   J. T. Robinson 1956, The Dentition of the Australopithecinae, *Transv. Mus. Mem.*, **9** : 110 (TM 1515); 48 (Sts 50).   *Plesianthropus transvaalensis*.

14.   ————

15.   J. T. Robinson 1956 (Sts 50).

16.   R. Broom 1941, The Origin of Man, *Nature, Lond.*, **148** : 10–14.   G. H. R. von Koenigswald 1948, Remarks on the lower canine of *Plesianthropus transvaalensis* Broom, *in* A. L. du Toit (ed.) 1948, Robert Broom Commemorative Volume, *Roy. Soc. S. Afr.*, Cape Town: 159–164.

  1.   Sterkfontein, ' Type Site ' (1947–1949 mandibulae).

  3.   R. Broom and J. T. Robinson, 24 June 1947 (Sts 7), 26 September 1947 (Sts 18), 11 March 1948 (Sts 24), 10 August 1948 (Sts 36), 1948 (Sts 38, Sts 41); J. T. Robinson, 1949 (Sts 62).

11.   **Sts 7.**   old adult, male: mandibula (f), crushed, with complete dentition except for rt C (lost in life); lt ascending ramus; rt scapula (acromion) (f); proximal portion of rt humerus (5 pieces) associated.

        Sts 18.   juvenile: mandibula (f) with lt dm2 (f), M1; rt dm1 (f), dm2, M1 (M2) (f).

        Sts 24.   juvenile, probably male: mandibula (f) with lt di1, di2, dc, dm1, (I1), (P3), (P4) (f); rt di1 (f), dc, dm1, dm2, M1, (I1), (I2), (P3).

        Sts 36.   adult, probably male: mandibula (f), distorted, in three pieces, lt I1 (f), C-M3; rt M2, M3.

        Sts 38.   old adult: lt corpus mandibulae (ff) with M1 (ff), M2 (ff), isolated M3.

        Sts 41.   mandibula (ff) with lt M3.

        Sts 62.   mandibula (ff) with alveoli of 2 I1–2; very worn lt C, P3 root (f).

12.   R. Broom and J. T. Robinson 1947, Further Remains of the Sterkfontein Ape-Man, *Plesianthropus, Nature, Lond.*, **160** : 430 (Sts 7).   R. Broom and J. T. Robinson 1949, A new mandible of the ape-man *Plesianthropus transvalensis, Am. J. phys. Anthrop.*, **7** : 123–127 (Sts 36).   Transvaal Museum Catalogue (unpublished) (Sts 38, Sts 62).

13.   R. Broom, J. T. Robinson and G. W. H. Schepers 1950, Sterkfontein Ape-Man *Plesianthropus, Transv. Mus. Mem.*, **4** : 34–36, 45–47, 56–57 (Sts 7) : 42–43, 50–55 (Sts 24).   J. T. Robinson 1952, Some hominid features of the ape-man dentition, *J. dent. Ass. S. Afr.*, **7** : 102–113 (Sts 36).   J. T. Robinson 1956, The Dentition of the

Australopithecinae, *Transv. Mus. Mem.,* **9** : 48, 110, 114 (Sts 7): 104, 141 (Sts 18): 36, 37, 104, 130, 132, 136 (Sts 24): 114 (Sts 41).   *Plesianthropus transvaalensis.*

14. ————

15. R. Broom, *et al.* 1950 (Sts 7, Sts 24).   J. T. Robinson 1956 (Sts 24).   R. A. Dart 1962, A Cleft Adult Mandible and the Nine other Lower Jaw Fragments from Makapansgat, *Am. J. phys. Anthrop.,* **20** : 267–286 (Sts 36).

1. Sterkfontein, (1971 mandibula).

3. Z. Mweli, E. Gabocoe, A. R. Hughes and P. V. Tobias, 15 November 1971 (StW/H14).

4. In breccia east of type locality and SE of the East Pit, 9 m ESE of position from which Sts 5 was recovered, and 30 cm deeper than StW/H 13.   (Grid Q/30, about 4 m below datum wire or approximately 2·75 m below ground datum).   P. V. Tobias 1972, Summary Progress Report on Palaeo-Anthropological Programme during 1971 (Departmental Report).

7. Association with artifacts not established.

11. **StW/H 14.**   young adult: crushed lt (f) and rt corpora and part of rt ramus mandibulae, with lt P4, M1 (f), M2 (f), M3 (f); rt P3, M1, M2, M3.

12. P. V. Tobias 1973, Darwin's prediction and the African emergence of the genus *Homo, in* L'Origine dell'Uomo, *Accademia Nazionale dei Lincei,* **182** : 63–85, Rome.

13. P. V. Tobias 1974, A new chapter in the history of the Sterkfontein early hominid site, *J. S. Afr. biol. Soc.,* **14** : 30–44.

14. ————

15. P. V. Tobias 1973.   P. V. Tobias 1974.

1. Sterkfontein, dump derived from Member 4 (1973–1976 mandibula).

11. **StW/H 39.**

1. Sterkfontein, ' Type Site ' (1938–1949 dentes maxillares).

3. R. Broom, 1938 (TM 1524, TM 1561, Sts 21).   R. Broom and J. T. Robinson, 11 July 1947 (Sts 11), 12 August 1947 (Sts 16), August 1947 (Sts 22), 28 November 1947 (Sts 23), 20 May 1948 (Sts 28), 11 June 1948 (Sts 30), 5 June 1948 (Sts 31), 28 May 1948 (Sts 33), July 1948 (Sts 37), 1948 (Sts 39, Sts 44, Sts 45, Sts 46), January 1949 (Sts 47), 1949 (Sts 48, Sts 54, Sts 55, Sts 56), 1947–1949 (Sts 72).

11. All upper dentes:

TM 1524.   very worn M1 (roots).

TM 1527.   rt C.

TM 1532.   lt M3, very worn (f).

TM 1561.   rt M3 crown (f).

Sts 11.	lt M1, M2. (Specimens missing).
Sts 16.	M, very worn (f).
Sts 21.	upper rt M crown (probably M1).
Sts 22.	lt M1, M2.
Sts 23.	rt M2 (f).
Sts 28.	rt M2, M3, lt M1 (Same individual as Sts 37).
Sts 30.	P4, rt M2.
Sts 31.	lt M3 (f).
Sts 33.	P (f).
Sts 37.	lt M2, M3 (same individual as Sts 28).
Sts 39.	lt P4, M1 (f), very worn.
Sts 43.	lt M1.
Sts 44.	lt M2 (f), M3 (crushed).
Sts 45.	M (f) (rt M2?).
Sts 46.	M3 crown (f).
Sts 47.	P3.
Sts 48.	lt C.
Sts 54.	rt M3, lt P3, P4.
Sts 55.	lt P3.
Sts 56.	M1, M2, dm2 (f).
Sts 72.	rt M3 (f) crown.

12. R. Broom and G. W. H. Schepers 1946, The South African Fossil Ape-Men: the Australopithecinae, *Transv. Mus. Mem.*, **2** : 57, 61, 62, 64 (TM 1527, TM 1532, TM 1561, Sts 21).   R. Broom, J. T. Robinson and G. W. H. Schepers 1950, Sterkfontein Ape-Man *Plesianthropus, Transv. Mus. Mem.*, **4** : 40 (Sts 48).   Transvaal Museum Catalogue (unpublished) (TM 1524, Sts 11, Sts 16).

13. J. T. Robinson 1956, The Dentition of the Australopithecinae, *Transv. Mus. Mem.*, **9** : 44, 45 (TM 1527) 93 (TM 1571) 84 (Sts 21) 88 (Sts 22) 88, 93 (Sts 37) 58 (Sts 47) 44, 45 (Sts 48) 93 (Sts 54 Two teeth given as Sts 54 on page 58 and 62 have been renumbered.): 58 (Sts 55) 84, 88 (Sts 56).

14. ————

15. R. Broom, *et al.* 1950 (Sts 22, Sts 48).   J. T. Robinson 1956 (TM 1561, Sts 21, Sts 28, Sts 33, Sts 37, Sts 47, Sts 55).

1. Sterkfontein, (1971 dens maxillaris).

3. P. V. Tobias, 8 October 1971.

4. Square Q.31, 1 m from StW/H 13.   P. V. Tobias and A. R. Hughes, *Annual Reports on Sterkfontein Excavation for 1971,* Department of Anatomy, University of Witwatersrand.

11. **StW/H 15.**   upper rt I1 (f).

12. P. V. Tobias 1972, *Summary Progress Report on Palaeo-Anthropological Programme during 1971,* Department of Anatomy, University of Witwatersrand.

13. ————

14. ————

15. ————

1.　Sterkfontein, ' Type Site ' (1938–1949 dentes mandibulares).
3.　R. Broom, 1938 (TM 1518, TM 1519, TM 1520, TM 1523, TM 1528, TM 1534); R. Broom and J. T. Robinson, 10 April 1947 (Sts 3, Sts 4), 27 June 1947 (Sts 9), 1948 (Sts 40, Sts 43); J. T. Robinson, 1949 (Sts 49, Sts 51, Sts 55b, Sts 59).
11.　All lower dentes:

TM 1518.　rt M3.
TM 1519.　rt M3, much worn.
TM 1520.　lt M3, much worn.
TM 1523.　lt P4.
TM 1528.　rt C, crown broken, root intact.
Sts　3.　rt C.
Sts　4.　M2.
Sts　9.　rt M1, root system not yet developed.
Sts 40.　C (f).
Sts 49.　P3 (f) crown.
Sts 51.　rt P3, (C).
Sts 55b.　lt M3 crown.
Sts 59.　lt M3 crown.
TM 1534.　I, very worn (possibly not australopithecine).

12.　W. K. Gregory and M. Hellman 1939, The Dentition of the extinct South African Man-Ape *Australopithecus (Plesianthropus) transvaalensis* Broom, *Ann. Transv. Mus.,* **19** : 339–373 (TM 1518, TM 1523, TM 1528).　R. Broom and G. W. H. Schepers 1946, The South African Fossil Ape-Men: the Australopithecinae, *Transv. Mus. Mem.,* **2** : 67, 68, 71 (TM 1519, TM 1520, TM 1523, TM 1528).
13.　J. T. Robinson 1956, The Dentition of the Australopithecinae, *Transv. Mus. Mem.,* **9** : 48, 51, 71, 75, 104, 110, 114.
14.　————
15.　J. T. Robinson 1956.

1.　Sterkfontein (1972 maxilla).
3.　A. R. Hughes and P. V. Tobias, 25 May 1972.
4.　Limeworkers' dump 13 (derived from type locality or East Pit, Member 4).
11.　**StW/H 18.**　adult: rt maxilla (f), P3 (f), P4 (ff), M1 (worn) and isolated rt I, (f) (worn), M3?
12.　P. V. Tobias 1973, A new chapter in the history of the Sterkfontein early hominid site, *J. S. Afr. biol. Soc.,* **14** : 30–44.

1.　Sterkfontein (1968–1972 isolated dentes).
3.　Members of excavation team under A. R. Hughes and P. V. Tobias, 1968–1972; Elias Gaboecoe, A. R. Hughes and P. V. Tobias, 10 June 1968 (StW Hom 1); A. R. Hughes and P. V. Tobias, 25 September 1968 (StW/H 5); A. R. Hughes and P. V. Tobias, 25 March 1969 (StW/H 2); A. R. Hughes and P. V. Tobias, 3 July 1969

(StW/H 3); Zebulon Mweli, A. R. Hughes and P. V. Tobias, 8 August 1969 (StW/H 4); A. R. Hughes and P. V. Tobias, 23 September 1969 (StW/H 10); A. R. Hughes and P. V. Tobias, 25 November 1969 (StW/H 6); A. R. Hughes and P. V. Tobias, 6 January 1970 (StW/H 7); A. R. Hughes and P. V. Tobias, 25 May 1971 (StW/H 11); A. R. Hughes and P. V. Tobias, 3 June 1971 (StW/H 16); A. R. Hughes and P. V. Tobias, 23 August 1971 (StW/H 9A); A. R. Hughes and P. V. Tobias, 12 April 1972 (StW/H 9B).

4. Limeworkers' dumps 12, 13 and 15, probably derived from Member 4.   P. V. Tobias and A. R. Hughes, Annual Reports on Sterkfontein Excavation for 1969, 1970, 1971, 1972, *Department of Anatomy, University of Witwatersrand, Johannesburg.*

11. **StW/H 1.**   lower lt M1, moderately worn.   (Same individual as Sts 4).
    StW/H 2.   upper rt M3, unworn.
    StW/H 3.   lower lt M, probably M3.   Slightly worn.   Roots intact, incompletely formed.
    StW/H 4.   roots and crown (f) of upper lt M, probably M1 or M2.
    StW/H 5.   I, probably lower.
    StW/H 6.   upper lt M3, unworn, roots missing.
    StW/H 7.   lower lt P3, moderately worn.
    StW/H 9.   A: upper rt P4; B: upper rt M1 articulating with each other approximally, both with marked occlusal and approximal wear.
    StW/H 10.   fragment of crown of (?) rt upper P.
    StW/H 11.   fragment of worn crown M.
    StW/H 16.   very worn upper lt P.

12. P. V. Tobias 1972, Summary Progress Report on Palaeo-Anthropological Programme during 1971, *Department of Anatomy, University of Witwatersrand, Johannesburg.*   P. V. Tobias and A. R. Hughes 1972, Resumé of Research Work at Sterkfontein, January–June 1972 (Departmental Reports).

13. P. V. Tobias 1974, A new chapter in the history of the Sterkfontein early hominid site, *J. S. Afr. biol. Soc.* **14** : 30–44.

14. ————

15. ————

1. Sterkfontein, dumps derived from Member 4 (1973–1976 dentes).
11. **StW/H 20.**   C.
    StW/H 21.   C crown.
    StW/H 22.   C root.
    StW/H 23.   I in jaw fragment.
    StW/H 24.   I.
    StW/H 32.   I.
    StW/H 35.   P.
    StW/H 36.   M.

StW/H 37.   M.
StW/H 43.   M.
StW/H 44.   C.
StW/H 45.   P.
StW/H 47.   M.
StW/H 48.   M.
StW/H 50.   P.
StW/H 51.   M.
StW/H 52.   M.
StW/H 54.   M.
StW/H 55.   P.
StW/H 56.   P.
StW/H 57.   P.
StW/H 58.   C.
StW/H 60.   P.
StW/H 61.   M.
StW/H 62.   I.

1.   Sterkfontein (1969 and 1975) (articulated vertebrae).
2.   Probably from Sterkfontein cave breccia.
3.   P. V. Tobias, 1969.   (Excavated many years earlier and removed from R. M. Cooper's display cabinet.)   P. V. Tobias, 10 January 1975.
4.   Reddish-brown breccia (probably Member 4).   Dump 18 (StW/H 41).   F. Ankel and P. V. Tobias, New Hominoid Vertebrae from the fossil breccias in the Sterkfontein area, in preparation.
11.   **StW/H 8.**   adult: four articulated vertebrae lumbales (f).
        StW/H 41.   adult: two articulated vertebral bodies (T12 and L1 or T11 and T12). (Same individual as StW/H 8.)
12.   P. V. Tobias 1973, A new chapter in the history of the Sterkfontein early hominid site, *J. S. Afr. biol. Soc.,* **14** : 30–44.
13.   F. Ankel and P. V. Tobias, in preparation.
14.   ———
15.   F. Ankel and P. V. Tobias, in preparation.   P. V. Tobias 1973.

1.   Sterkfontein, ' Type Site ' (1936–1949 post-cranial bones).
3.   R. Broom, 1936 (TM 1513), 1938 (TM 1526).   R. Broom and J. T. Robinson, 1 August 1947 (Sts 14), 29 June 1948 (Sts 34).   J. T. Robinson, 1949 (Sts 65, Sts 68).
11.   **TM 1513.**   distal end of lt femur.
        TM 1526.   rt os capitatum.
        Sts 14.   adult female: 7 vertebrae thoracicales; 6 vertebrae lumbales; 2 centra vertebrae; 2 sacral vertebrae, 4 costae (f); rt and lt os coxae (f); diaphysis of lt femur (f), tibia? (ff).

Sts 34.	distal extremity of rt femur.
Sts 65.	rt os coxae (f) with most of ilium and part of pubis; proximal end of lt femur (ff – probably not hominid); centrum vertebrae (ff).
Sts 68.	proximal end of lt radius (probably baboon).
Sts 73.	centrum vertebrae.

12. R. Broom and G. W. H. Schepers 1946, The South African Fossil Ape-Men: the Australopithecinae, *Transv. Mus. Mem.*, **2** : 73–75 (TM 1513, TM 1526).   R. Broom and J. T. Robinson 1950, Notes on the Pelves of the Fossil Ape-Men, *Am. J. phys. Anthrop.*, **8** : 489–494 (Sts 14, Sts 65).

13. R. Broom and G. W. H. Schepers 1946 : 73–75 (TM 1513, TM 1526).   R. Broom, J. T. Robinson and G. W. H. Schepers 1950, Sterkfontein Ape-Man *Plesianthropus*, *Transv. Mus. Mem.*, **4** : 58–63 (Sts 14, Sts 65).   *Plesianthropus transvaalensis.*   R. Broom and J. T. Robinson 1949, The Lower End of the Femur of *Plesianthropus*, *Ann. Transv. Mus.*, **21** : 181–182 (Sts 34).

14. J. T. Robinson 1972, *Early Hominid Posture and Locomotion*, Chicago and London.

15. J. T. Robinson 1972.

16. W. E. Le Gros Clark 1947, Observations on the Anatomy of the Fossil Australopithecinae, *J. Anat.*, **81** : 324–330 (TM 1513, TM 1526).   W. E. Le Gros Clark 1955, The os innominatum of the recent Pongidae with special reference to that of the Australopithecinae, *Am. J. phys. Anthrop.*, **13** : 19–28.   J. R. Napier 1964, The Evolution of Bipedal Walking in the Hominids, *Arch. Biol. (Liège)*, **75** suppl. : 673–708.   A. Walker 1973, New *Australopithecus* femora from East Rudolf, Kenya, *J. hum. Evol.*, **2** : 545–555.

1. Sterkfontein, dumps derived from Member 4 (1973–1976 post-cranial bones).

11.	**StW/H 25.**	head of rt femur.
	StW/H 26.	rt 4th metacarpale.
	StW/H 28.	proximal phalanx manus.
	StW/H 29.	proximal phalanx manus.
	StW/H 30a.	head and neck of lt femur.
	StW/H 30b.	head of rt femur.
	StW/H 30c, d, e.	os coxae? (ff).
	StW/H 31.	head of rt femur.
	StW/H 38.	distal humerus (ff).
	StW/H 46.	distal end of lt radius.
	StW/H 63.	metacarpale.
	StW/H 64.	metacarpale.

MEMBER 5

1. Sterkfontein, ' Extension Site ' (1957–1958 maxilla, dentes).

3. J. T. Robinson, 11 April 1957 (SE 255), 1957 (SE 1508, SE 1579), 25 May 1958 (SE 1937), 29 September 1958 (SE 2396).

7. See general entry (p. 118).

11.   **SE   255.**   juvenile: maxilla (ff) with M1 crown, dm2, dm1 (f).
      SE 1508.   upper rt M2.
      SE 1579.   upper lt M1 or M2 (f).
      SE 1937.   lower lt C.
      SE 2396.   lingual half of upper lt P3.
12.   J. T. Robinson 1958, The Sterkfontein Tool-Maker, *The Leech,* **28** : 96.
13.   J. T. Robinson 1962 : 102 (SE 255, SE 1502, SE 1937).   *Australopithecus africanus.*
      P. V. Tobias 1965, *Australopithecus, Homo habilis,* tool-using and tool-making, *S. Afr.
      archaeol. Bull.,* **20** : 187 (SE 1579, SE 2396).   *Homo erectus ('Telanthropus')?*
14.   ————
15.   ————

 1.   Sterkfontein (1976 skull).
 3.   A. R. Hughes, 9–17 August 1976.
 4.   Partly in calcified and partly in decalcified breccia, attributed to Member 5, south of
      West Pit.   A. R. Hughes and P. V. Tobias 1977, A fossil skull probably of the Genus
      Homo from Sterkfontein, Transvaal, *Nature, Lond.,* **265** : 310–312.
11.   **StW/H 53.**   adult: much of calvaria, ossa faciei, part of basis cranii, palatinum, 9
                      dentes, ramus mandibulae (ff).
12.   Anon, 1976, Important fossil skull found at Sterkfontein, *S. Afr. J. Sci.,* **72** : 227.
13.   ————
14.   ————
15.   A. R. Hughes and P. V. Tobias 1977.
16.   ————
17.   Department of Anatomy, University of the Witwatersrand, Medical School, Hospital
      Street, Johannesburg.
18.   ————

 1.   Sterkfontein, dumps derived from Member 5 (1973–1976 dentes).
11.   **StW/H 19a.**   P.
      StW/H 19b.   M2, M3.
      StW/H 42.   I.

 1.   Sterkfontein, dumps derived from Member 5 (1973 post cranial).
11.   **StW/H 27.**   distal 1/3 of metacarpale.

SWARTKRANS

General Entry

1. Swartkrans.
2. Breccia-filling of ancient cave in dolomitic limestone, Sterkfontein valley about 1·6 km NW of Sterkfontein, 13 km NNW of Krugersdorp, Transvaal.   26° 01′ S, 27° 43′ E.
3. Members of staff of the Transvaal Museum, Pretoria.   R. Broom and J. T. Robinson, 1948, 1949; J. T. Robinson, 1952; C. K. Brain, 1965–1972.
4. Lower Breccia Phase II.   C. K. Brain 1958, The Transvaal Ape-Man-Bearing Cave Deposits, *Transv. Mus. Mem.,* **11** : 78–80.   Member 1, pink breccia (*Paranthropus robustus* remains and *Homo* cranium SK846/847/80); Member 2, brown breccia (SK15).   C. K. Brain 1976, A re-interpretation of the Swartkrans site and its remains, *S. Afr. J. Sci.,* **72** : 141–146.
5. ————
6. Early Middle Pleistocene.   K. P. Oakley 1957, Dating the Australopithecines, *Proc. 3rd Pan-Afr. Congr. Prehist.* : 155–157.   Final stages of First Interpluvial.   C. K. Brain 1958 : 88.   Swartkrans Faunal Span.   H. B. S. Cooke 1967, The Pleistocene Sequence in South Africa and Problems of Correlation, *in* W. W. Bishop and J. D. Clark (eds) 1967, *Background to Evolution in Africa,* Chicago and London : 175–184.   Swartkrans Faunal Span (Member 1); Cornelia Span (Member 2).   E. S. Vrba 1975, Some Evidence of Chronology and Palaeoecology of Sterkfontein, Swartkrans and Kromdraai from the fossil Bovidae, *Nature, Lond.,* **254** : 301–304.
7. Pebble tools. cf. Oldowan.   C. K. Brain 1958 : 88.   Developed Oldowan.   M. D. Leakey 1970, Stone artefacts from Swartkrans, *Nature, Lond.,* **225** : 1222–1225.
8. *Papio robinsoni, Simopithecus danieli, Tapinochoerus meadowsi, Phacochoerus antiquus, Lycyaena silberbergi, Crocuta crocuta angella.*   R. F. Ewer 1957, Faunal evidence on the dating of the Australopithecinae, *Proc. 3rd Pan-Afr. Congr. Prehist.* : 135–142.
   SKa (Member 1): *Damaliscus* or *Parmularius,* cf. *Connochaetes taurinus,* cf. *Megalotragus, Redunca* cf. *arundinum, Pelea* cf. *capreolus, Antidorcas* cf. *recki, A. bondi,* cf. *Gazella vanhoepeni, Oreotragus* cf. *major, Syncerus, Tragelaphus* cf. *scriptus, T.* cf. *strepsiceros, T.* sp. aff. *angasi,* cf. *Makapania.*
   SKa (Member 2): *Damaliscus* cf. *dorcas, D. ?niro,* cf. *Connochaetes taurinus, Hippotragus* cf. *niger,* cf. *Kobus ellipsiprymnus, Pelea* cf. *capreolus, Antidorcas* cf. *australis* and *A.* cf. *marsupialis, A. bondi, Oreotragus* cf. *oreotragus, O.* cf. *major, Raphicerus* cf. *campestris, Ourebia* cf. *ourebi, Tragelaphus* cf. *scriptus, T.* cf. *strepsiceros, Taurotragus* cf. *oryx.*   E. S. Vrba 1975.   C. K. Brain, E. S. Vrba & J. T. Robinson 1974, A new hominid innominate bone from Swartkrans, *Ann. Transv. Mus.,* **29** : 55–63.   E. S. Vrba 1976, The Fossil Bovidae of Sterkfontein, Swartkrans and Kromdraai, *Transv. Mus. Mem.,* **21** : 1–166.
9. ————
10. A3: > 160,000 BP on the basis of protoactinium/thorium dating of underlying stalagmite.   J. N. Rosholt and P. S. Antal 1963, Evaluation of the Pa^{231}/U-Th^{230}U method for dating Pleistocene Carbonate rocks, *U.S. Geol. Surv. Prof. Pap.,* **450-E** : E108–111.
    A4: < 2·57 Myr on basis of geomorphological dating of cave opening.   T. C. Partridge

1973, Geomorphological Dating of Cave Openings at Makapansgat, Sterkfontein, Swartkrans and Taung, *Nature, Lond.,* **246** : 75–79.

11.–15.   See under individual entries.

16.   J. T. Robinson 1954, The Genera and Species of the Australopithecinae, *Am. J. phys. Anthrop.,* **12** : 181–200. *Paranthropus robustus crassidens.*   W. E. Le Gros Clark 1964, *The Fossil Evidence for Human Evolution,* 2nd edn, Chicago : 123–173. *Australopithecus robustus.*   J. T. Robinson 1958, Cranial cresting patterns and their significance in the Hominoidea, *Am. J. phys. Anthrop.,* **16** : 397–428.   J. T. Robinson 1972, *Early Hominid Posture and Locomotion* : 1–361, Chicago and London.   C. K. Brain 1973, The significance of Swartkrans, *J. S. Afr. biol. Soc.,* **13** : 13–17.   P. V. Tobias 1973, Implications of the New Age Estimates of the Early South African Hominids, *Nature, Lond.,* **246** : 79–83.   H. M. McHenry 1974, How large were the Australopithecines?, *Am. J. phys. Anthrop.,* **40** : 329–340 (SK 18b, 82, 97).   A. E. Mann 1975, Some paleodemographic aspects of the South African australopithecines, *Univ. Pennsylv. Publs. Anthrop.,* **1** : 1–171.

17.   Transvaal Museum, P.O. Box 413, Pretoria, Transvaal.

18.   While stocks last casts of the most important specimens can be bought from the Wenner-Gren Foundation, 14 East 71st Street, New York, N.Y. 10021, U.S.A.   (SK 12, SK 13, SK 18, SK 45, SK 48, SK 50, SK 61, SK 64, SK 82, SK 84, SK 85, SK 97, SK 847e, SK 3155b).

Index of Fossil Hominid Remains from Swartkrans

Reg. No.	Element	Page	Reg. No.	Element	Page
SK 1	M2	141	SK 24	P3	143
SK 2	I1	137	SK 25	mandibula	139
SK 3/40	rt I1	137	SK 27	cranium	138
SK 4	C	137	SK 28	P4	143
SK 5	M2	141	SK 29	C	141
SK 6	mandibula	137	SK 30	P3	141
SK 7	P4	141	SK 31	M3	143
SK 9	P4	141	SK 32	P4	143
SK 10	mandibula	139	SK 33	P3	141
SK 11	maxilla	140	SK 34	mandibula	139
SK 12	mandibula	139	SK 35	M1	143
SK 13	maxilla	140	SK 36	M3	143
SK 14	maxilla	140	SK 37	mandibula	139
SK 15	mandibula	148	SK 38	C	143
SK 16	M3	143	SK 39	P4	143
SK 17	M1	143	SK 40/3	I1 I	143
SK 18	P, radius	148	SK 41	M3	143
SK 19	M	141	SK 42	M2	143
SK 20	M1	141	SK 43	P4	148
SK 21	maxilla	140	SK 44	P3	143
SK 22	M3	141	SK 45	mandibula	139
SK 23	mandibula	140	SK 46	cranium	138

Reg. No.	Element	Page	Reg. No.	Element	Page
SK 47	cranium	138	SK 105	M3	143
SK 48	cranium	138	SK 438	mandibula	139
SK 49	cranium	138	SK 820	C	142
SK 50	os coxae	144	SK 821	ossa faciei, P3	143
SK 52	cranium	138	SK 822	P3	143
SK 54	calotte	138	SK 823	P3	143
SK 55	maxilla	140	SK 824	P4	143
SK 57	maxilla	140	SK 825	P4	143
SK 61	mandibula	139	SK 826a/877	maxilla	140
SK 62	mandibula	139	SK 826b	P4	142
SK 63	mandibula	139	SK 827	P4	142
SK 64	mandibula	139	SK 828	M1	142
SK 65/74c	maxilla	140	SK 829	M1	143
SK 66	maxilla	140	SK 830	P4	142
SK 67	I1	140	SK 831	P3	142
SK 68	I1	143	SK 831a	maxilla	140
SK 69	I1	143	SK 832	M1	143
SK 70	I1	143	SK 833	M1	143
SK 71	I2	143	SK 834	M2	143
SK 72	P3	141	SK 835	M3	143
SK 73	I1	143	SK 836	M3	143
SK 74a	mandibula	139	SK 837	M2	143
SK 74b	I3	141	SK 838a	maxilla	140
SK 74c	P4	143	SK 838b	M1	142
SK 75	M3	141	SK 839	maxilla	140
SK 79	cranium	138	SK 840	M3	142
SK 80	maxilla	138	SK 841a	mandibula	140
SK 81	P4, M1	139	SK 841b	M3	142
SK 82	femur	144	SK 842/869	dm2	140
SK 83	cranium	138	SK 843	mandibula	140
SK 84	metacarpale	147	SK 844	mandibula	140
SK 85	metacarpale	147	SK 845	maxilla	140
SK 85a	C	143	SK 846a	M1	142
SK 86	C	143	SK 846b	cranium	138
SK 87	C	141	SK 847	cranium	138
SK 88	P4	141	SK 848	cranium	138
SK 89	M1	143	SK 849	M1	143
SK 90	dm2	143	SK 850	P3	141
SK 91	dm1	143	SK 851	M3	142
SK 92	C	143	SK 852	mandibula	140
SK 93	C	143	SK 853	vertebra	144
SK 94	C	143	SK 854	vertebra	144
SK 95	C	143	SK 855	M3	142
SK 96	mandibula	139	SK 856	P4, M1/2	143
SK 97	femur	144	SK 857	P3	142
SK 98	M2	143	SK 858	mandibula	140
SK 99	P4	143	SK 859	cranium	138
SK 100	P3	141	SK 860	humerus	144
SK 101	P3	143	SK 861	mandibula	140
SK 102	M1	143	SK 862	mandibula	140
SK 104	M1	141	SK 863	M	143

Reg. No.	Element	Page	Reg. No.	Element	Page
SK 864	M1?	142	SK 1591	M1	145
SK 865	M	142	SK 1592	maxilla	145
SK 866	dentes	142	SK 1593	P3	145
SK 867	P3	143	SK 1594	M1	145
SK 868	M2, M3	143	SK 1595	maxilla	145
SK 869/842	mandibula	140	SK 1596	C	145
SK 870	M3	143	SK 1648	mandibula	145
SK 871	M3	142	SK 3155b	ilium	146
SK 872	M1	143	SK 3974	M1	145
SK 873	I	142	SK 3975	M3	145
SK 874	dens	142	SK 3976	M2	145
SK 875	dens	142	SK 3977	M3	145
SK 876	mandibula	140	SK 3978	mandibula	145
SK 877	maxilla	140	SK 3981	vertebra	146
SK 878	dentes	142	SK 14000	M	145
SK 879	dentes	142	SK 14001	P3	145
SK 880a, b	M3	142	SK 14003	cranium	145
SK 881	maxilla	140	SK 14079	P	145
SK 882	M	140	SK 14080	maxilla	145
SK 883	M3	142	SK 14128	P3	145
SK 884	C, mastoideus	143	SK 14129a	maxilla	145
SK 885	M3	142	SKW 5	mandibula	145
SK 1512	palatinum	145	SKW 6	M	145
SK 1514	mandibula	145	SKW 7	dentes	145
SK 1524	M3	145	SKW 8	2M	145
SK 1585	endocast	145	SKW 10	M	145
SK 1586	mandibula	145	SKW 11	maxilla	145
SK 1587	mandibula	145	SKW 12	maxilla	145
SK 1588	mandibula	145	SKW 14	M	145
SK 1589	P4	145	SKW 15	M	145
SK 1590	palatinum, mandibula femur head, pelvis	145, 147	SKW 27	metacarpale	146

MEMBER 1

1. Swartkrans (1948 adolescent).
3. R. Broom and J. T. Robinson, 16–26 November 1948.
11. **SK 6.**   adolescent: mandibula (f) with lt P3, P4, M1, M2, (M3); isolated rt P4, M1, M2, M3.   **Holotype** of *Paranthropus crassidens* Broom, 1949.   Plate 11.   (Same individual as SK 100, and probably as SK 13/14).
 SK 2.   upper rt I1.
 SK 3.   upper rt I1 (same individual as SK 40).
 SK 4.   upper rt C.
12. R. Broom 1949, Another new type of fossil Ape-Man, *Nature, Lond.,* **163** : 57.
13. R. Broom and J. T. Robinson 1952, Swartkrans Ape-Man, *Transv. Mus. Mem.,* **6.**
 J. T. Robinson 1956, The Dentition of the Australopithecinae, *Transv. Mus. Mem.,*

9 : 69, 73, 101, 112 (SK 6) : 23 (SK 2) : 26 (SK 3) : 41 (SK 4). *Paranthropus crassidens.*

14. ————

15. R. Broom and J. T. Robinson 1952.   J. T. Robinson 1956.

1. Swartkrans (1949–1952 crania).
3. R. Broom and J. T. Robinson, 1949–1952; quarryman Fourie, 30 June 1950 (SK 48).
11. **SK 27.**   juvenile, probably male: crushed cranium (f) with most of braincase but ossa faciei damaged and incomplete; maxillae (f) with lt I2, dm2, (P4) M1; rt M1; unerupted C, P3, M2 removed.

    SK 46.   adult, probably female: cranium (ff) crushed, with lt half of braincase, portion of face, palatinum and maxilla with lt P4, M1, M2, M3; rt P3—M3.

    SK 47.   adolescent, possibly female: cranium (f) much crushed, with most of base and palatinum; lt M1 (ff); rt (P4), M1, M2, (M3).

    SK 48.   adult, probably female: cranium (f), partly crushed, with damage to base and occipitale, palatinum, upper lt P4 (roots) M1, M2, M3, rt C, P3, P4 (roots), M1–3 (roots) (same individual as SKW 7).

    SK 49.   adult: cranium (f), much crushed, with part of parieto-occipitale separate, upper lt and rt P3, (f, f), P4, M1, M2, M3.

    SK 52.   adolescent, probably male: cranium (ff) with lower part of face and portion of right side, upper lt I2 (f), P3, P4, M1 (ff), rt I2, P3, P4, M1, (M3).

    SK 54.   juvenile: calotte (f), crushed.

    SK 79.   adult: anterior portion of crushed cranium (f) with much of face and palatium, lt P3, P4, M1, M2, M3; rt P3 (ff), P4, M1, M2, M3.

    SK 83.   adult: major portion of severely damaged cranium (f), palatinum (f), 2 I1–2, (roots), lt C, P3, P4, M1, M2, M3; rt M3.

    SK 846b. left meatus acusticus externus, petrous and mastoid parts of temporale, (same individual as SK 80/SK 847 and most probably SK 45).

    SK 847. elderly adult: cranium (ff), with left M3, (same individual as SK 80/846b).

    SK 80.   maxilla (ff) with lt I2, P3 (f) and 2 isolated dentes (same individual as SK 847/846b and most probably SK 45).

    SK 848. auditory region (f) showing meatus acusticus externus and glenoid fossa.

    SK 859. juvenile: major portion of occipitale (f), parietale (f).

12. R. Broom and J. T. Robinson 1952, Swartkrans Ape-Man, *Transv. Mus. Mem.,* **6** : 26–28 (SK 27): 14–15 (SK 46): 28–30 (SK 47): 10–13 (SK 48).   J. T. Robinson 1953a, The Nature of *Telanthropus capensis, Nature, Lond.,* **171** : 33 (SK 80).   J. T. Robinson 1954a, The Australopithecine Occiput, *Nature, Lond.,* **174** : 262–3 (SK 49).   R. J. Clarke, F. Clark Howell and C. K. Brain 1970, More evidence of an advanced hominid at Swartkrans, *Nature, Lond.,* **225** : 1219–1222 (SK 846b, SK 847).   *Homo* sp.

13. R. Broom and J. T. Robinsom 1952.   J. T. Robinson 1954b, Nuchal Crests in the Australopithecines, *Nature, Lond.,* **174** : 1197.   J. T. Robinson 1956, The Dentition of the Australopithecinae, *Transv. Mus. Mem.,* **9** : 41, 55, 81, 86, 90 (SK 27, SK 48); 60, 81, 86, 90, 91 (SK 46, SK 49); 26, 55, 60, 81 (SK 47, SK 52); 60 (SK 79); 41 (SK 83).   *Paranthropus crassidens.*   Unpublished (SK 54, SK 848, SK 859).

14. ————
15. R. Broom and J. T. Robinson 1952 (SK 48).   J. T. Robinson 1953b (SK 45, SK 80).   J. T. Robinson 1956 (SK 48, SK 49).   J. T. Robinson 1958, Cranial Cresting Patterns and their Significance in the Hominoidea, *Am. J. phys. Anthrop.*, **16** : 403–408 (SK 49).   R. J. Clarke, F. Clarke Howell and C. K. Brain 1970 (SK 846b, SK 847).

1. Swartkrans (1949 mandibulae).
3. R. Broom and J. T. Robinson, 1949 (SK 10, SK 25, SK 34, SK 37, SK 63, SK 64, SK 74A, SK 96, SK 438); J. T. Robinson, 1 April 1949 (SK 12); July 1949 (SK 61); Nov. (SK 62).
11. **SK 10.** adult: lt corpus mandibulae (f) with M2 (f), M3 (ff) (probably same individual as SK 11).
SK 12. elderly male: mandibula (f) with lt and rt P3 (roots), P4 (f), M1, M2, M3, palatinum (f), maxillae (f), with lt I2 (root), P3 (roots), P4, M1 (roots); rt P3, P4, M1 (f), M2 (ff).
SK 25. juvenile, probably male: mandibula with lt P4 (erupting), M1, M2, rt P4, M1, M2 (probably same individual as SK 832).
SK 34. adult, probably male: mandibula in two halves, with lt P4, M1, M2, M3, rt I1, I2, C, P3, P4, M1, M2, M3.
SK 37. adult: mandibula (ff) with lt M2, and small portion of lt M1.
SK 45. elderly adult: rt corpus mandibulae (f) with M1, M2, alveolus of M3 (probably same individual as SK 80/847/846b).
SK 61. juvenile: mandibula (ff) with rt and lt di1, di2 (f), dc, dm1, dm2, rt (M1).
SK 62. juvenile: mandibula (ff) with lt di2, dc (f), dm1, dm2, (M1), rt I1, dc, dm1 (f), dm2.
SK 63. juvenile: mandibula with lt and rt dc, dm1, dm2, M1, (M2), rt di1, (C) (probably same individual as SK 89/90/91).
SK 64. juvenile: mandibula (ff) with rt dm1, dm2.
SK 74a. adult female: mandibula (f) showing possible mental eminence, with lt P3 (ff), P4 (ff), M1 (ff), M2 (ff), rt I2 (f), P3, P4, M1 (f), M2.
SK 81. lt P4, M1.
SK 96. juvenile: mandibula (ff) with roots lt dm1, unerupted lt C, P3.
SK 438. juvenile: mandibula (ff) with lt (dm2).
12. R. Broom and J. T. Robinson 1950a, Man contemporaneous with the Swartkrans Ape-Man, *Am. J. phys. Anthrop.*, **8** : 151–156 (SK 45).   *Homo* sp.   R. Broom and J. T. Robinson 1950b, Note on the skull of the Swartkrans Ape-Man *Paranthropus crassidens, Am. J. phys. Anthrop.*, **8** : 295–300 (SK 12, SK 45, SK 74a).   R. Broom and J. T. Robinson 1952, Swartkrans Ape-Man, *Transv. Mus. Mem.*, **6** : 22 (SK 25): 16–18 (SK 34).   J. T. Robinson 1954: 189 (SK 96).
13. R. Broom and J. T. Robinson 1952: 4–9 (SK 12); 16–18 (SK 34).   J. T. Robinson 1953, The Nature of *Telanthropus capensis, Nature, Lond.*, **171** : 33 (SK 46).   J. T. Robinson 1956, The Dentition of the Australopithecinae, *Transv. Mus. Mem.*, **9** : 60, 73, 112 (SK 12): 73, 101, 107 (SK 25): 35, 36, 46, 69, 73, 101, 107, 112 (SK 34, SK 74a): 107 (SK 37): 73, 112 (SK 81): 46, 69 (SK 96): 101, 130, 131, 134, 134, 140 (SK 61): 35, 130, 131, 134, 140 (SK 62): 35, 37, 46, 101, 131, 134, 140 (SK 63): 134, 140 (SK 64).   Transvaal Museum Catalogue (unpublished) (SK 10, SK 438).

14. —————
15. R. Broom and J. T. Robinson 1950 (SK 12, SK 74a). R. Broom and J. T. Robinson 1952 (SK 12).   J. T. Robinson 1956 (SK 25, SK 96).   R. A. Dart 1962, A Cleft Adult Mandible and the Nine Other Lower Jaw Fragments from Makapansgat, *Am. J. phys. Anthrop.*, **20** : 267–286 (SK 34).

1. Swartkrans (1950–1952 mandibulae).
3. Quarryman Fourie, 1950 (SK 23); J. T. Robinson, 1952 (SK 841–876).
11. **SK 23.**   adult, probably female: mandibula with dentes.
   SK 841a.   juvenile: mandibula (ff) with lt dm2, M1 (f) (not related to 841b).
   SK 843.   adolescent: lt corpus mandibulae (f) with M1, M2 (M3) (same individual as SK 846a, and SK 75 and, probably, as SK 14129a/SK 105).
   SK 844.   adult: corpus mandibulae and lt ramus (f) with M2 (f), M3.
   SK 852.   juvenile: mandibula (f) with lt dm1 (f), dm2, rt dm1, dm2 (f), M1 (f) (most probably same individual as SK 839).
   SK 858.   adult: mandibula (f) with lt I1, I2 (f), C (root), P3, P4, M1 (f), M2 (ff), rt I1, I2, C, P3 (f), P4, M1, M2, M3 (f), all considerably cracked.
   SK 861.   mandibula (ff) with lt M1 (ff), M2 (ff).   (Same individual as SK 858 & 883.)
   SK 862.   mandibula (ff) with rt P4 (roots), M1 (roots), M2 (f), M3 (f).
   SK 869.   juvenile: mandibula (ff) with broken roots of dm1 (same individual as SK 842).
   SK 876.   mandibula (ff), crushed, with lt C root, P3 (f), P4–M3, rt I2 (f), C-M3.
12. R. Broom and J. T. Robinson 1952, Swartkrans Ape-Man, *Transv. Mus. Mem.*, **6** : 19–21 (SK 23).   J. T. Robinson 1954, The Genera and Species of the Australopithecinae, *Am. J. phys. Anthrop.*, **12** : 196 (SK 841–876). *Paranthropus robustus crassidens.*
13. J. T. Robinson 1956, The Dentition of the Australopithecinae, *Transv. Mus. Mem.*, **9** : 35, 36, 46, 69, 73, 101, 107, 112 (SK 23): 140 (SK 841a); 101, 107 (SK 843); 112 (SK 844); 35, 36, 101, 107 (SK 858–incorrectly given as SK 845). *Paranthropus robustus crassidens.*
14. T. D. White 1976, *Anns. Transv. Mus.*, **30** : 97–98 (SK 858, 861).
15. J. T. Robinson 1956 (SK 23, SK 841a, SK 843).

1. Swartkrans (1949–52 maxillae).
3. R. Broom and J. T. Robinson 1949 (SK 21, SK 54, SK 55, SK 57, SK 65–67); February 1949 (SK 11); April 1949 (SK 13, SK 14); J. T. Robinson, 1952 (SK 826a–882).
11. **SK 11.**   adult: face with lt P3, P4, M1, M2, M3, rt P3 (f), P4 (f), M1 (f), M2 (probably same individual as SK 10).
   SK 13/14. adolescent: face with upper lt P3, P4, M1, M2 occlusal caries, (M3), rt P3, P4, M1, M2, (M3) (probably same individual as SK 6/100).
   SK 21.   maxilla (ff) with lt C (ff), P3, P4, M1, M2, M3.
   SK 55/55b. juvenile: maxilla with lt I1, I2, C, P3, dm2, M1, M2 (f), rt I1, I2, C, P3,

dm2; mandibula (f) with lt P3, dm2, M1, M2 (f), rt dm2 (f), M1, M2, (M3).

SK 57.	palatinum (ff) with lt P3, P4, M1, M2, M3 (f), rt P3, P4 (ff), M1 (f).
SK 65.	adult: maxilla (f) with lt I1, I2, C, P3, P4, isolated rt I1, C (same individual as SK 67 and SK 74C).
SK 66.	juvenile: rt maxilla (f) with roots of dc, dm1, dm2, incompletely developed crowns of I1, I2.
SK 67.	rt I1 (probably same individual as SK 65/74c).
SK 826a.	adult: lt maxilla (ff) with P3 (f), P4, M1 (f), M2, (possibly same individual as SK 877, not related to SK 826b).
SK 831a.	adult: maxilla (ff) with lt M2, M3 (not related to SK 831).
SK 838a.	juvenile: maxilla (ff) with rt dm2, M1 (not related to SK 838b, most probably same individual as SK 102).
SK 839.	juvenile: maxilla (ff) with damaged lt and rt di, dm2, M1 (probably same individual as SK 852).
SK 845.	adult: crushed face with lt I2 (f), C, P3, P4, M1, M2 (f), rt P3, P4, M1 (ff).
SK 877.	maxilla (f) with rt P4, M1, M2 (possibly same individual as SK 826a).
SK 881.	maxilla (f) with lt P3, P4, M1 (ff) (possibly same individual as SK 882).
SK 882.	upper lt M (f) (possibly same individual as SK 881).

12. J. T. Robinson 1956, The Dentition of the Australopithecinae, *Transv. Mus. Mem.,* **9** : 60, 86, 91 (SK 11): 55, 60, 81, 86, 90 (SK 13): 23, 26, 41, 55, 81 (SK 55): 69, 101, 112, 140 (SK 55b): 55, 60, (SK 57): 23, 26, 41, 55, 60 (SK 65): 26 (SK 66): 86 (SK 826a): 86, 91 (SK 831a): 81, 128 (SK 838a): 81, 122, 128, (SK 839): 55, 60, 107 (SK 845 – specimens given as SK 845 on pages 35, 36, 101, 107 have been renumbered SK 858). Transvaal Museum Catalogue (unpublished) (SK 21, SK 877, SK 881, SK 882). *Paranthropus robustus crassidens.*

13. J. T. Robinson 1956.

14. ———————

15. J. T. Robinson 1952, Some hominid features of the Ape-Man dentition, *J. dent. Ass. S. Afr.,* March 15 (SK 13). J. T. Robinson 1956 (SK 13, SK 14, SK 65, SK 826a, SK 831a, SK 839).

1. Swartkrans (1949 dentes mandibulares).

3. R. Broom and J. T. Robinson, 1949.

11.

SK	**1.**	lt M2.
SK	**5.**	lt M2.
SK	**7.**	rt P4 ⎫ (most probably same individual).
SK	**9.**	lt P4. ⎭
SK	**19.**	rt M2.
SK	**20.**	lt M1.
SK	**22.**	rt M3 (probably same individual as SK 880a).
SK	**29.**	rt C.
SK	**30.**	lt P3.
SK	**33.**	rt P3.
SK	**72.**	lt P3.

      SK 74b.   rt I1 (not related to 74a or 74c).
      SK 75.   rt (M3) (same individual as SK 843/846a and, most probably SK 1412-9a/SK 105.)
      SK 87.   rt C.
      SK 88.   lt P4.
      SK 100.   rt P3 (same individual as SK 6 and SK 13/14).
      SK 104.   rt (M1).
      SK 850.   rt P3 (f).

12. J. T. Robinson 1956, The Genera and Species of the Australopithecinae, *Am. J. phys. Anthrop.*, **12** : 189 (SK 87).   *Paranthropus robustus crassidens.*

13. J. T. Robinson 1956, The Dentition of the Australopithecinae, *Transv. Mus. Mem.*, **9** : 107 (SK 1) : 73 (SK 7, SK 9) : 112 (SK 22) : 46, 48 (SK 29) : 69 (SK 30, SK 72, SK 100) : 35 (SK 74b) : 73, 112 (SK 81) : 46 (SK 87) : 73 (SK 88 – incorrectly given as SK 81) : 101 (SK 104).   *Paranthropus robustus crassidens.*

14. ————

15. J. T. Robinson 1956 (SK 1, SK 9, SK 22, SK 29, SK 72, SK 87, SK 94, SK 100).

  1. Swartkrans (1952 dentes mandibulares).
  3. J. T. Robinson, 1952.
11. **SK 820.**   C.
      SK 826b.   lt P4 (not related to 826a).
      SK 827.   lt P4.
      SK 828.   lt M1.
      SK 830.   lt P4.
      SK 831.   rt P3 (not related to 831a).
      SK 838b.   lt M1 (not related to 838a).
      SK 840.   lt M3 (possibly same individual as SK 855).
      SK 841b.   lt M3. (not related to 841a).
      SK 842.   lt dm2 (same individual as SK 869).
      SK 846a.   rt M1 (not related to SK 846b; same individual as SK 843/75 and, most probably, as SK 826a/SK 14129a/SK 105).
      SK 851.   rt M3.
      SK 855.   rt M3 (most probably same individual as SK 840).
      SK 857.   rt P3.
      SK 864.   M1? (f).
      SK 871.   lt M3.
      SK 880
        a & b.   lt M3 associated with 2 very broken dentes, (probably same individual as SK 22).
      SK 883.   rt M3 and associated angle of mandibula (f).   (Same individual as SK 858, 861.)
      SK 885.   lt M3.
      Positions indeterminable:
      SK 865.   worn, crushed M with fragments.
      SK 866.   dentes (ff) in breccia.
      SK 873.   I (f).

  SK 874. dens (f).
  SK 875. dens (f).
  SK 878. 2 dentes (f) and small fragment of calotte.
  SK 879. fragments of dentes and associated bone fragments.
12. J. T. Robinson 1956, The Genera and Species of the Australopithecinae, *Am. J. phys. Anthrop.,* **12** : 1–179. *Paranthropus robustus crassidens.*
13. J. T. Robinson 1956, The Dentition of the Australopithecinae, *Transv. Mus. Mem.,* **9** : 46 (SK 820) : 73 (SK 826b, SK 830) : 69 (SK 831) : 101 (SK 828, SK 838b, SK 846) : 69–71 (SK 857 incorrectly listed as SK 827) : 112 (SK 840, SK 841b), 140 (SK 842). *Paranthropus robustus crassidens.*
14. ————
15. J. T. Robinson 1956.

 1. Swartkrans (1949–1952 dentes maxillares).
 3. R. Broom and J. T. Robinson, 1949; J. T. Robinson, 1952.
11. **SK 16.** lt M3.
  SK 17. rt M1 (probably same individual as SK 1591).
  SK 24. lt P3.
  SK 28. lt P4.
  SK 31. rt M3.
  SK 32. rt P4.
  SK 35. lt M1 (f).
  SK 36. rt M3.
  SK 38. rt C.
  SK 39. rt P4 (f).
  SK 40. lt I1 (same individual as SK 3).
  SK 41. lt M3.
  SK 42. M2 (f).
  SK 44. rt P3 (abnormal).
  SK 68. lt I1.
  SK 69. lt I1 crown (same individual as SK 73).
  SK 70. lt I1.
  SK 71. rt (I2).
  SK 73. rt (I1) (probably same individual as SK 69).
  SK 74c. rt P4 (not related to 74a, 74b; same individual as SK 65/67).
  SK 85a. rt C.
  SK 86. lt C (f).
  SK 89/90/91. lt and rt M1, lt dm2, rt dm1 (probably same individual as SK 63).
  SK 92. rt (C).
  SK 93. lt C (probably same individual as SK 85a).
  SK 94. lt C.
  SK 95. lt C.
  SK 98. lt M2.
  SK 99. lt P4.
  SK 101. lt P3.
  SK 102. lt M1 (most probably same individual as SK 838a).

SK 105.	lt M3 (same individual as SK 826a/SK 14129a and most probably SK 843/846a/75).
SK 821.	portion of face with lt P3.
SK 822.	lt P3.
SK 823.	rt P3.
SK 824.	lt P4.
SK 825.	lt P4.
SK 829.	lt M1.
SK 832.	lt M1 (probably same individual as SK 25).
SK 833.	lt M1.
SK 834.	rt M2.
SK 835.	lt M3.
SK 836.	lt M3 (f).
SK 837.	rt M2, M3.
SK 849.	rt M1.
SK 856.	rt P4, (M1 or 2).
SK 863.	M crown (f).
SK 867.	P3 (f).
SK 868.	rt M2 (f), M3 (f).
SK 870.	lt M3 (f).
SK 872.	lt M1.
SK 884.	lt C and associated mastoid region.

12. J. T. Robinson 1956.
13. J. T. Robinson 1956, The Dentition of the Australopithecinae, *Transv. Mus. Mem.*, **9** : 23 (SK 40, SK 68, SK 69, SK 73) : 41 (SK 38, SK 92, SK 93, SK 95) : 26 (SK 70, SK 71) : 46, 48 (SK 86, listed as a lower canine SK 94) : 55 (SK 24, SK 33) : 51 (SK 95) : 60 (SK 28, SK 32, SK 99) : 81 (SK 17, SK 102) : 91 (SK 16, SK 31, SK 36, SK 41) : 86 (SK 98) : 81, 123, 128 (SK 89, SK 90, SK 91) : 55 (SK 821, SK 822, SK 823) : 60 (SK 824, SK 825) : 81 (SK 829, SK 832) : 86 (SK 834, SK 837) : 91 (SK 835, SK 836). *Paranthropus robustus crassidens.*
14. ————
15. J. T. Robinson 1956.

1. Swartkrans (1949, 1952 post-cranial bones).
3. R. Broom and J. T. Robinson, 1949 (SK 50, SK 82, SK 97); J. T. Robinson, 1952 (SK 853, SK 854, SK 860).
11. **SK 50.** adult, probably male: major portion of rt os coxae, pubis and crista iliaca missing.
    SK 82. mature adult: proximal portion of rt femur.
    SK 97. mature adult: proximal portion of rt femur.
    SK 853. juvenile: vertebra lumbalis.
    SK 854. axis.   (Probably not hominid.)
    SK 860. distal end of rt humerus, somewhat crushed (associated with SK 50).
12. R. Broom and J. T. Robinson 1950, Notes on the Pelves of the Fossil Ape-Men, *Am. J. phys. Anthrop.*, **8** : 489–494 (SK 50).

13. R. Broom and J. T. Robinson 1952, Swartkrans Ape-Man, *Transv. Mus. Mem.*, **6** : 90–94. Transvaal Museum Catalogue (unpublished) (SK 853, SK 854, SK 860). *Paranthropus crassidens.*
14. J. T. Robinson 1972, *Early Hominid Posture and Locomotion* : 1–361, Chicago and London.
15. J. T. Robinson 1972, figs 49–52, 64, 71–73.
16. W. E. Le Gros Clark 1950, New Discoveries of the Australopithecinae, *Nature, Lond.*, **166** : 758. W. E. Le Gros Clark 1955, The os innominatum of the recent Ponginae with special reference to that of the Australopithecinae, *Am. J. phys. Anthrop.*, **13** : 19–28. J. R. Napier 1964, The Evolution of Bipedal Walking in the Hominids, *Archs Biol., Liège*, **75** suppl. : 673–708 (SK 50). A. Walker 1973, New *Australopithecus* femora from East Rudolf, Kenya, *J. hum. Evol.*, **2** : 545–555.

1. Swartkrans (1948–1952, miscellaneous specimens).
3. Fossils retrieved by Museum staff during 1966 from breccia excavated by R. Broom and J. T. Robinson from 1948–1952 and stored in the Transvaal Museum.
11. **SK 1512.**   palatinum (ff), roots of dentes.
    SK 1514.   mandibula (ff), M1 (roots), M2 (ff).
    SK 1524.   upper lt (M3) (ff).
    SK 1591.   upper lt M1 (probably same individual as SK 17).
    SK 1592.   palatinum (f) with rt P4, M1, M2, M3 (ff).
    SK 1595.   maxillary fragment with upper lt dm 2, fragments of erupting teeth.
    SK 1596.   upper C.
    SK 14003. part of a crushed skull (not prepared): lt M1 (f), rt M1 (f), M2, M3.
12. C. K. Brain 1970, New Finds at the Swartkrans australopithecine site, *Nature, Lond.*, **225** : 1112–1119. *Paranthropus robustus.*
13. ————
14. ————
15. ————

1. Swartkrans (1965–72, crania and dentes).
2. C. K. Brain and Staff of the Transvaal Museum, in miners' dumps of cave breccia, April 1965 (SK 1590), September 1965 (SK 1586), January 1966 (SK 1585), March 1966 (SK 1593), July 1966 (SK 1587), September 1966 (SK 1594), October 1966 (SK 1588, SK 1589), November 1966 (SK 1648), May 1967 (SK 3976, SK 3977), July 1967 (SK 3974, SK 3975, SK 14001), August 1967 (SK 3978), November 1967 (SK 14000), July 1969 (SK 14079, SK 14080), September 1969 (SK 14128), January 1970 (SK 14129a), July 1970 (SKW 14, SKW 15), October 1970 (SKW 5a), November 1970 (SKW 6), December 1970 (SKW 7), January 1971 (SKW 5b), February 1971 (SKW 8), September 1971 (SKW 10), Mrs E. A. Voigt, March 1972 (SKW 11); C. K. Brain in miners' dumps of cave breccia, March 1972 (SKW 12).
11. **SK 1585.**   rt half of endo-cranial cast.

L

SK 1586.	mandibula (ff) with rt I1, I2, M1 (ff), M2, M3, lt I1, P (ff), M1, M2, M3.
SK 1587.	lt mandibula (f), P3 (roots), P4, M1, M2, isolated rt M2.
SK 1588.	rt mandibula (ff), P3 (roots), P4, M1, M2 (roots ff).
SK 1589.	rt (P4).
SK 1590.	palatinum (ff) with rt I2, C, P3, P4, M1, mandibula (ff) with dentes (ff), femur head (ff). (See also p. 147).
SK 1593.	lower rt P3.
SK 1594.	lower rt M1 (ff).
SK 1648.	crushed mandibula, lt P4 (roots), M1, (f), M2 (roots); rt M1 (f), M2.
SK 3974.	lower rt M1 crown.
SK 3975.	upper lt M3.
SK 3976.	lower lt M2.
SK 3977.	upper rt M3.
SK 3978.	infant: corpus mandibulae, rt and lt dm1, dm2, lt (M1).
SK 14000.	upper M (f).
SK 14001.	upper lt P3.
SK 14079.	P (f).
SK 14080.	maxilla (ff), lt M2 (part of SK 877).
SK 14128.	upper rt P3.
SK 14129a.	maxilla fragment with rt M1, M2 (same individual as SK 826a/105 and, most probably SK 843/846a/75).
SKW 5.	mandibula.
SKW 6.	M crown (ff).
SKW 7.	crowns of upper rt P4, M1, M2. (fits cranium SK 48).
SKW 8.	2 upper M.
SKW 10.	M (f).
SKW 11.	maxilla (f).
SKW 12.	lt maxilla with P3–M3.
SKW 14.	M crown (f).
SKW 15.	M crown (ff).

12. C. K. Brain 1970, New finds at the Swartkrans australopithecine site, *Nature, Lond.*, **225** : 1112–1119 (SK 1585–SK 1590, SK 1593–4, SK 1648, SK 3974–3978, SK 14000–1).   C. K. Brain 1967, The Transvaal Museum's fossil Project at Swartkrans, *S. Afr. J. Sci.*, **63** : 378–384.   Less robust hominid than *Paranthropus* (SK 1587 and SK 1588).

13. R. L. Holloway 1972, New Australopithecine endocast, SK 1585, from Swartkrans, South Africa, *Am. J. phys. Anthrop.*, **37** : 173–186 (SK 1585).

14. ─────────

15. R. L. Holloway 1972 (SK 1585).   C. K. Brain 1967 (SK 1587, SK 1588, SK 1585).

16. R. L. Holloway 1973, Endocranial volumes of early African Hominids, and the role of the brain in human mosaic evolution, *J. hum. Evol.*, **2** : 449–459.

1.   Swartkrans (1949 ossa metacarpalia).
3.   R. Broom and J. T. Robinson, 1949.
11.   **SK 84.**   metacarpale I (C. K. Brain 1966, *in lit.)* but SK 84 used for upper canine *in* J.
            T. Robinson 1956 : 41.
      SK 85.   metacarpale IV.   This number has also been given to an upper canine of 1949
            (see page 143).
12.   R. Broom and J. T. Robinson 1949, Thumb of the Swartkrans Ape-Man, *Nature,*
      Lond., **164** : 841–2 (SK 84).
13.   J. R. Napier 1959, Fossil Metacarpals from Swartkrans, *Fossil Mammals Afr.* No. 17,
      British Museum (Natural History) London. *Paranthropus crassidens* (SK 84).
      Probably *Homo erectus ('Telanthropus')* (SK 85).
14.   J. T. Robinson 1972, *Early Hominid Posture and Locomotion* : 1–361, Chicago and
      London: 197–206.
15.   J. T. Robinson 1972, figs 101–102.

1.   Swartkrans (1968–72 post-cranial bones).
3.   C. K. Brain and the Staff of the Transvaal Museum in miners' dumps of cave breccia,
      April 1965 (SK 1590), March 1968 (SK 3981); in Primary breccia stored in the Trans-
      vaal Museum, previously excavated by R. Broom and J. T. Robinson between 1948 and
      1952, retrieved 1970 (SK 3155b); *in situ* at base of main excavation face, February
      1972 (SKW 27).
11.   **SK 1590.**   fragments of os coxae.   (See also p. 145).
      SK 3981.   vertebra thoracicalis (last), vertebra lumbalis (last) (f).
      SK 3155b.   rt ilium (i) with complete acetabulum.
      SKW 27.   adult: lt metacarpale V.
12.   C. K. Brain 1970, New Finds at the Swartkrans australopithecine site, *Nature, Lond.,*
      225 : 1112–1119 (SK 3981).   M. H. Day and J. L. Scheuer 1973, SKW 14147: A new
      hominid metacarpal from Swartkrans, *J. hum. Evol.,* **2** : 429–438 (SKW 27).   C. K.
      Brain, E. S. Vrba and J. T. Robinson 1974, A new hominid innominate bone from
      Swartkrans, *Anns Transv. Mus.,* **29** : 55–63 (SK 3155b).   *Homo erectus.*
13.   J. T. Robinson 1970, Two new early Hominid Vertebrae from Swartkrans, *Nature,*
      Lond., **225** : 1217–1219 (SK 3981).   H. M. McHenry 1975, A new pelvic fragment
      from Swartkrans and the relationship between the robust and gracile australopithecines,
      Am. J. phys. Anthrop., **43** : 245–261 (SK 3155b).
14.   J. T. Robinson 1972, *Early Hominid Posture and Locomotion,* London and
      Chicago : 113–120 (SK 3981).
15.   J. T. Robinson 1972 figs 64–69 (SK 3981).   M. H. Day and J. L. Scheuer 1973 (SKW
      27).   C. K. Brain *et al.* 1974 (SK 3155b).
16.   H. M. McHenry and R. S. Corruccini 1975, Multivariate analysis of early hominid
      pelvic bones, *Am. J. phys. Anthrop.,* **43** : 263–270.   H. M. McHenry 1975, Fossils and
      the Mosaic Nature of Human Evolution, *Science N.Y.,* **190** : 425–431 (SK 3155b).

MEMBER 2

1. Swartkrans (1949).
3. J. T. Robinson, 29 April 1949 (SK 15, SK 18, SK 43).
4. Lower Breccia pocket refilled with dark brown consolidated deposit.   R. Broom and J. T. Robinson 1952, Swartkrans Ape-Man, *Transv. Mus. Mem.,* **6** : 110, 113.   Material in pocket identical with that of Lower Breccia Phase II (' pink breccia '), except that carbonate content exceptionally low.   C. K. Brain 1958, The Transvaal Ape-Man-Bearing Cave Deposits, *Transv. Mus. Mem.,* **13** : 87 (SK 15, SK 18, SK 43).
9. Sk 15 dentine : $F = 0.7\%$, $100F/P205 = 2.8$.
   Sk 15 bone : $F = 0.37\%$, $100F/P205 = 1.5$, $eU308 = 3$ ppm, $N = $ nil.
   Animal bone : $eU308 = 5$ ppm.
11. **SK 15.**   mature adult: mandibula (f) with lt M1, M2, M3, rt M2, M3.   **Holotype** of *Telanthropus capensis* Broom & Robinson, 1949.   Plate 13.
    SK 18.   lower lt P3 and associated proximal end of lt radius.
    SK 43.   lower rt P4 buccal crown (f).
    All three probably from same individual.
12. R. Broom and J. T. Robinson 1949, A new type of fossil man, *Nature, Lond.,* **164** : 322–323 (SK 15, SK 18, SK 43).
13. R. Broom and J. T. Robinson 1952 : 110–111, 113–115.   J. T. Robinson 1953b, *Telanthropus* and its phylogenetic significance, *Am. J. phys. Anthrop.,* **11** : 462–477.   *Telanthropus capensis* (SK 15, SK 18, SK 43).   R. Broom and J. T. Robinson 1952 : 111–112.
14. ─────────
15. R. Broom and J. T. Robinson 1952.
16. A. Simonetta 1957, Catalogo e sinonimia annotata degli ominoidi fossili ed attuali, 1758–1955, *Atti soc. Tosc. Sci. Nat.,* **64** : 51–112.   *Homo erectus.*   J. T. Robinson 1961, The Australopithecines and their bearing on the origin of man and of toolmaking, *S. Afr. J. Sci.,* **57** : 3–13.   *Homo erectus* (SK 15, SK 18, SK 43).   W. E. Le Gros Clark 1964, *The Fossil Evidence for Human Evolution,* 2nd edn, Chicago : 127–129, 156, 170.   *Australopithecus.*

TAUNG

1. Taung (formerly Taungs).
2. Fissure deposit in travertine, at limeworks of Northern Lime Company, Buxton, 9·6 km SW of Taung, Cape Province.   27° 32′ S, 24° 48′ E.
3. M. de Bruyn, 1924.
4. Pink calcified sand in limestone, 15 m below surface.   K. P. Oakley 1954, Dating of the Australopithecinae of South Africa, *Am. J. phys. Anthrop.,* **12** : 11.   Cave deposit of red sandy limestone.   F. E. Peabody 1954, Travertines and cave deposits of the Kaap escarpment of South Africa, and the type locality of *Australopithecus africanus* Dart, *Bull. geol. Soc. Am.,* **65** : 671–706.   Pink calcified clay (42%) and sand (8%) in limestone, 15 m below surface, containing the hominid fossil; and an underlying red calcified sand (94%) and clay (3%), containing the baboon fossils.   K. W. Butzer 1974, Paleoecology of South African Australopithecines: Taung Revisited, *Curr, Anthrop.,* **15** : 367–382.
5. ─────────

6. Not older than Lower Pleistocene, possibly Middle Pleistocene.   F. E. Peabody 1954 : 705.   Late Villafranchian.   K. P. Oakley 1957, Dating the Australopithecines, *Proc. 3rd Pan-Afr. Congr. Prehist.* : 155–157.   Sterkfontein Faunal Span.   H. B. S. Cooke 1968, Evolution of Mammals on Southern Continents. II. The Fossil Mammals of Africa, *Quart. Rev. Biol.*, **43** (3) : 234–264.   Between Sterkfontein and Swartkrans Faunal Spans.   L. H. Wells 1971, Africa and the Ancestry of Man, *S. Afr. J. Sci.*, **67** (4) : 276–283.   The australopithecine level is younger than the baboon deposit⁓ and compatible with Peabody's (1954) Norlim travertine phase; the hominid clearly is no older than the youngest time-range represented at Swartkrans or Kromdraai.   K. W. Butzer 1973.

7. None associated.

8. *Parapapio antiquus, Papio izodi, Crocidura taungsensis.*   R. F. Ewer 1957, *Proc. 3rd Pan-Afr. Congr. Prehist.* : 136.   *Papio wellsi* (cf. *P. angusticeps* from Kromdraai).   L. Freedman 1961, New Cercopithecoid fossils, including a new species, from Taung, Cape Province, South Africa, *Ann. S. Afr. Mus.*, **46** : 1–14.   *Parapapio jonesi, P. whitei, Cercopithecoides williamsi.*   L. Freedman 1970, A new check-list of fossil Cercopithecoidea of South Africa, *Palaeont. afr.*, **13** : 109–110.   All probably belong to the older, red sand, and are thus not contemporaneous with the hominid.

9. Hominid mandible: F = 0·3%, 100F/P205 = 3·8.   Blesbuck bone (coll. 1924): F = 0·4%, 100F/P205 = 1·4.

10. The earliest date for opening of the Taung fissure is < 0·87 Myr.   T. C. Partridge 1973, Geomorphological dating of cave openings at Makapansgat, Sterkfontein, Swartkrans and Taung, *Nature, Lond.*, **246** : 75–79.

11. **Taung 1.**   child, 5–6 years: calotte (ff) (with natural endocast), basis cranii, ossa faciei, mandibula, upper and lower dentition with lt and rt di1, di2, dc, dm1, dm2, M1.   **Holotype** of *Australopithecus africanus* Dart, 1925.   Plate 14.

12. R. A. Dart 1925, *Australopithecus africanus*, the man-ape of South Africa, *Nature, Lond.*, **115** : 195–199.

13. R. A. Dart 1929, A Note on the Taungs Skull, *S. Afr. J. Sci.*, **26** : 648–658.   R. A. Dart 1934, The Dentition of *Australopithecus africanus, Folia anat. jap.*, **12** : 207–221.   J. T. Robinson 1956, The Dentition of the Australopithecinae, *Transv. Mus. Mem.*, **9** : 85, 107, 122, 126, 129, 133, 137, 143.

14. ————

15. R. Broom and G. W. H. Schepers 1946, The South African Fossil Ape-Men: the Australopithecinae, *Transv. Mus. Mem.*, **2** : 1–272.   J. T. Robinson 1956.

16. J. T. Robinson 1954, The Genera and Species of the Australopithecinae, *Am. J. phys. Anthrop.*, **12** : 196.   *Australopithecus africanus africanus.*   W. E. Le Gros Clark 1964, *The Fossil Evidence for Human Evolution*, 2nd edn, Chicago : 123–173.   J. T. Robinson 1965, *Homo 'habilis'* and the Australopithecines, *Nature, Lond.*, **205** : 121–124.   R. L. Holloway 1970, Australopithecine endocast (Taung specimen 1924): A new volume determination, *Science, N.Y.*, **168** : 966–968.   P. V. Tobias 1971, *The Brain in Hominid Evolution*, New York and London, 170 pp.   R. L. Holloway 1973, Endocranial Volumes of Early African Hominids, and the Role of the Brain in Human Mosaic Evolution, *J. hum. Evol.*, **2** : 449–459.   P. V. Tobias 1973, Implications of the New Age Estimates of the Early South African Hominids, *Nature, Lond.*, **246** : 79–83.   Possibly *Australopithecus robustus.*

17. Department of Anatomy, University of the Witwatersrand, Johannesburg, Medical School, Hospital Street, Johannesburg, Transvaal.

18. While stocks last, Wenner-Gren Foundation for Anthropological Research, 14 East 71st Street, New York, N.Y. 10021, U.S.A.

TSITSIKAMA 34° 10′ S, 24° 25′ E.

In a cliff-shelter at Tsitsikama W of Port Elizabeth, Cape Province, F. W. Fitzsimons discovered, in 1922, fossil hominid remains at two levels:

A. 'at great depth', 6 crania (f) and post-cranial bones of 4 or 5 individuals (f).   R. A. Dart 1923, Boskop remains from SE Africa Coast, *Nature, Lond.,* **112** : 623–625.   H. S. Gear 1926, A further report on the Boskopoid remains from Zitzikama, *S. Afr. J. Sci.,* **23** : 923–934.

B. Associated with kitchen midden industry, one skeleton, 7 crania; 'Strandlooper'.   G. D. Laing 1924, A preliminary report on some Strandlooper Skulls found at Zitzikama, *S. Afr. J. Sci.,* **21** : 528–541.

The hominid material is property of the Port Elizabeth Museum, Port Elizabeth, Cape Province.

TUINPLAAS

1. Tuinplaas (Springbok Flats).
2. Quarry, about 130 km N of Pretoria, Transvaal.   25° 00′ S, 28° 36′ E.
3. C. J. Swierstra and H. Lang, 1929.
4. Lateritic deposit on calcareous tufa.   H. Lang 1929, The Discovery of the ' Springbok ' Man, *Ill. Lond. News,* 16 March 1929 : 427–428.
5. Probably a contemporaneous burial (but may be slightly later) from Second Intermediate (Magosian) level.   L. H. Wells 1959, The Problem of Middle Stone Age Man in Southern Africa, *Man,* **59** : 158–160.
6. Late Pleistocene/Holocene.
7. Middle Stone Age (Pietersburg).   C. van Riet Lowe 1929, Notes on some Stone Implements from Tuinplaats, Springbok Flats, *S. Afr. J. Sci.,* **26** : 623–630.   R. J. Mason 1962, *Prehistory of the Transvaal,* Johannesburg : 233–235.
8. *Homoioceras [Bubalus] baini* at level of skeleton.   H. Lang 1929.
9. Skull: F = 0·6%, 100F/P205 = 2·7, N = 0·6%. Associated *Connochaetes* bone: F = 0·5%, 100F/P205 = 3·4, N = 0·3%.
10. A2: > 5500 BP on basis of C14 dating of calcite from calcareous crust on bone from fossil site dated as 5570 ± 100 BP (Pta-256).   The bone itself contains no collagen and may be considerably earlier.   J. C. Vogel and M. Marais 1971, Pretoria Radiocarbon Dates I, *Radiocarbon,* **13** (2) : 378–394.
11. **Tuinplaas 1.**   adult: cranium (ff), mandibula and post-cranial bones.
12. R. Broom 1929a, The Transvaal Fossil Human Skeleton, *Nature, Lond.,* **123** : 415–416.

13. G. W. H. Schepers 1941, The Mandible of the Transvaal Fossil Human Skeleton from Springbok Flats, *Ann. Transv. Mus.*, **20** : 253–271.   M. J. Toerien and A. R. Hughes 1955, The Limb Bones of Springbok Flats Man, *S. Afr. J. Sci.*, **52** : 125–128.

14. ————

15. R. Broom 1929b, The ' Bushveld ' Fossil (also called ' Springbok ' Man), *Ill. Lond. News,* 16 March 1929 : 426–427.   G. W. H. Schepers 1941.

16. A. Galloway 1937, Man in Africa in the Light of Recent Discoveries, *S. Afr. J. Sci.,* **34** : 89–120.   L. H. Wells 1952, Human Crania of the Middle Stone Age in South Africa, *Proc. 1st Pan-Afr. Congr. Prehist.* : 125–133.

17. Transvaal Museum, P.O. Box 413, Pretoria, Transvaal.

18. ————

SOUTH WEST AFRICA

Phillip Vallentine TOBIAS

University of the Witwatersrand, Johannesburg,
Medical School,
Hospital Street,
Johannesburg,
Transvaal.

&

Hertha de VILLIERS

University of the Witwatersrand, Johannesburg,
School of Dentistry,
Jan Smuts Avenue,
Johannesburg,
Transvaal.

OTJISEVA

1.   Otjiseva.
2.   Open site in small valley in the upper reaches of the Khomas Highlands, on the farm Otjiseva No. 45, 40 km NW of Windhoek.   22° 30′ S, 17° 00′ E.
3.   Farm worker and W. Sydow, 1964; W. Sydow and B. J. Grobbelaar, 1966.
4.   Terrace deposit of micaceous silt.   M. Martin 1970, *in* M. Sydow 1970, Report on some fossil human remains from Otjiseva, near Windhoek, South West Africa, *Scientific Research in South West Africa*, 9th Series : 1–48.
5.   ————
6.   Holocene.
7.   None reported.
8.   None reported.
9.   Otjiseva 1: eU308 $= 9$ ppm.   W. Sydow 1970.   N $= 0 \cdot 02\%$ (UCLA-1749).
10.   A1: $> 4440 \pm 70$ BP on basis of C14 dating of carbonate crust on bone (GrN-5347). W. Sydow 1970.
11.   **Otjiseva 1.**   adult male(?): calotte (ff), mandibula (f) with 3M, post-cranial bones (ff).
12.   W. Sydow 1969, Discovery of a Boskop skull at Otjiseva, near Windhoek, S.W.A., *S. Afr. J. Sci.*, **65** : 77–82.
13.   T. Jones 1970, *in* W. Sydow 1970.
14.   H. de Villiers 1972, The first Fossil Human Skeleton from South West Africa, *Trans. Roy. Soc. S. Afr.*, **40** (3) : 187–196.
15.   H. de Villiers 1972.
16.   ————
17.   State Museum, Windhoek.
18.   ————

SUDAN

Gordon Winant HEWES

&

David Lee GREENE

Department of Anthropology,
University of Colorado,
Boulder,
Colorado,
U.S.A.

&

Fred WENDORF

Department of Anthropology,
Southern Methodist University,
Dallas, Texas, 75222,
U.S.A.

KHARTOUM

1. Khartoum.
2. Open site on left bank of Blue Nile, NE of Khartoum Central Railway Station, under Eastern extension of Khartoum Civil Hospital.   15° 33′ N, 32° 33′ E.
3. F. Debono under the direction of A. J. Arkell, December 1944.
4. Calcified sand.   A. J. Arkell 1949, *Early Khartoum*, Oxford : 9–14.
5. Shallow burials.   A. J. Arkell 1949 : 34.
6. Holocene.
7. Khartoum Mesolithic including bone harpoons, pottery, ostrich eggshell beads and red ochre.   A. J. Arkell 1949 : 37–97.   A. J. Arkell 1953, *Shaheinab*, Oxford : 102.
8. Fauna includes *Thryonomys arkelli;* Aves; Reptilia; Pisces: *Synodontis schall;* Mollusca.   A. J. Arkell 1949 : 15–30.
9. ————
10. A3: *c.* 8000 BP on basis of C14 dating of shell associated with harpoon of Khartoum Mesolithic type from Tagra, N of Khartoum.   A. J. Arkell 1973, *in lit.*
11. **Khartoum 1–17.**   17 fragmentary burials including one complete skeleton (M20/2).
12. D. E. Derry 1949, *in* A. J. Arkell 1949 : 31–34.
13. D. E. Derry 1949.   ‘ Negroid ’.
14. ————
15. A. J. Arkell 1949 : pl. 7 (M20/2).
16. ————
17. Khartoum Antiquities Museum, P.O. Box 178, Khartoum.
18. ————

SAHABA   21° 45′ N, 31° 12′ E.

At an open site (117), on the E bank of the Nile, 3 km S of Wadi Halfa, R. Paepe and J. Guichard, under the direction of F. Wendorf excavated burials in a river terrace and removed fragmentary skeletons of 49 adults and 11 children.   50% of these skeletons were pierced by stone points of the Qadan industry.   The burials are stated by Wendorf to be probably about 12,000 BP on basis of associated artifacts.

F. Wendorf (ed.) 1968, *The Prehistory of Nubia*, Dallas: 33.

The remains are preserved in the Department of Antiquities, Khartoum (3 individuals) and the Department of Anthropology, Southern Methodist University, Dallas, Texas, 75222, U.S.A.

SINGA

1. Singa.
2. Eroded from caliche deposit exposed in the west bank of the Blue Nile, probably at low

water, about 46 m downstream of the District Commissioner's house at Singa, approximately 320 km S of Khartoum.   13° 00′ N, 33° 55′ E.

3.   W. R. G. Bond, February 1924.
4.   Limestone concretion in Gezira Clay.   G. W. Grabham 1938, Note of the Geology of
     the Singa district of the Blue Nile, *in* A. S. Woodward 1938, A Fossil Skull of an
     Ancestral Bushman from the Anglo-Egyptian Sudan, *Antiquity,* **12** : 193–195.
5.   ————
6.   Upper Pleistocene.   D. M. A. Bate 1951, The Mammals from Singa and Abu Hugar, *in*
     A. J. Arkell, D. M. A. Bate, L. H. Wells and A. D. Lacaille 1951, The Pleistocene
     Fauna of two Blue Nile Sites, *Fossil Mammals of Africa,* No. 2, British Museum (Na-
     tural History), London.   A. J. Arkell 1949, The Old Stone Age in the Anglo-Egyptian
     Sudan, *Sudan Antiquities Service,* Occasional Paper **1** : 45–48.
7.   Advanced Levalloisian or Proto-Stillbay.   A. D. Lacaille 1951, The Stone Age In-
     dustry of Singa—Abu Hugar, *in* A. J. Arkell *et al.* 1951.
8.   *Hystrix astasobae, Homoioceras singae, Crocodylus niloticus.*   D. M. A. Bate 1951, *in*
     A. J. Arkell *et al.* 1951.
9.   Human skull: F = 0·7%, 100F/P205 = 6·5, eU308 = 24 ppm, N = < 0·1%. Abu
     Hugar bones: F = 1·8%, 100F/P205 = 6·3, eU308 = 63 ppm, N = nil.
10. A3: *c.* 17,000 BP on basis of C14 dating on fossil crocodile tooth from Abu Hugar:
     17,300 ± 200 BP (I–712).
11. **Singa 1.**   calvaria (ff) (heavily mineralized).
12. A. S. Woodward 1938.
13. A. S. Woodward 1938.
14. L. H. Wells 1951, The Fossil Human Skull from Singa, *in* A. J. Arkell *et al.*
     1951.   Pre-Bushman.
15. A. S. Woodward 1938.   L. H. Wells 1951, *in* A. J. Arkell *et al.* 1951.
16. P. V. Tobias 1968, Middle and Early Upper Pleistocene Members of the Genus *Homo*
     in Africa, *in* G. Kurth (ed.), *Evolution und Hominisation,* 2 edn. Stuttgart : 188.
17. British Museum (Natural History), Cromwell Road, London SW7 5BD,
     England.   Reg. No. M 15546.
18. British Museum (Natural History), Cromwell Road, London SW7 5BD, England.

SOLEB

1.   Soleb.
2.   Open site, 400 m W of Nile, 300 m N of Soleb temple.   20° 26′ N, 30° 20′ E.
3.   Mission Michela Schiff Giorgini, 1959.
4.   Fossil bones scattered on the surface of the ground, carried by the Wadi water possibly
     from the upper terraces.   M. S. Giorgini (ed.) 1971, *Soleb II, les Nécropoles,* Firenze:
     392–393.
5.   ————
6.   Late Pleistocene.   Y. Coppens 1971, *in* M. S. Giorgini (ed.) 1971 : 393.
7.   Flint and small cornelian artifacts of late Levallois tradition eroded from same terrace at
     adjacent site.   J. de Heinzelin 1971, *in* M. S. Giorgini (ed.) 1971 : 394–395.
8.   Associated collection of fossil Vertebrate fragments: Mammalia, Reptilia and
     Pisces.   Y. Coppens 1971, *in* M. S. Giorgini (ed.) 1971 : 393–394.

9.   ————
10. ————
11. **Soleb 1.** adult: mandibula (ff).
     Soleb 2. adult: symphysis mandibulae (f).
     Soleb 3. adult: cranium (ff).
12. M. S. Giorgini 1959, Soleb, Campagna, 1958–9, *Kush,* **7** : 166.   Y. Coppens 1971, *in* M. S. Giorgini (ed.) 1971 : 393–394. *Homo sapiens.*
13. F. Sausse, in press, *W. Afr. J. Archaeol.,* Ibadan. *Homo sapiens.*
14. ————
15. Y. Coppens 1971, *in* M. S. Giorgini (ed.) 1971 : 393.   F. Sausse 1974.
16. ————
17. Temporarily Laboratoire d'Anthropologie du Muséum National d'Histoire Naturelle Musée de l'Homme, 75116 Paris, France.
18. Institut de Paléontologie du Muséum National d'Histoire Naturelle, 8 rue de Buffon, 75005 Paris, France.

WADI HALFA

1. Wadi Halfa.   (6B28).
2. Wind-eroded surface of occupation site 6B28, 140 m above sea level, 2·75 km N of Wadi Halfa, 1 km from left bank of Nile.   21° 57′ N, 31° 20′ E.
3. W. R. G. Bond, 1924.
4. Wind-eroded desert surface.   G. J. Armelagos 1964, A Fossilized Mandible from near Wadi Halfa, Sudan, *Man,* **64** : 12–13.   H. T. Irwin, J. B. Wheat and L. F. Irwin 1968, University of Colorado Investigations of paleolithic and epi-paleolithic sites in the Sudan, Africa, *Univ. Utah Anthrop. Paps,* No. 90.
6. Late Pleistocene.
7. Stone industry at site 6B28 included flakes in a derived Levalloisian technique and small backed blades.   G. J. Armelagos 1964.   H. T. Irwin *et al.* 1968.
8. Mainly bovids, one leporid.   G. J. Armelagos 1964.
9. Human mandible: F = 1·3%, 100F/P205 = 5·6, eU308 = 10 ppm, N = nil.   Bovid bone: F = 2·0%, 100F/P205 = 6·1, eU308 = 16 ppm, N = nil.
10. A3: *c.* 15,000 BP on basis of C14 dating from site 6B29 (GX 120).
11. **Wadi Halfa 6B28 1.**   adult: mandibula (ff), 12 dentes much eroded.
12. G. J. Armelagos 1964.
13. ————
14. ————
15. G. J. Armelagos 1964.
16. Compare Mesolithic burials from site 6B36.   G. W. Hewes, H. Irwin, M. Papworth and A. Saxe 1964, A New Fossil Human Population from the Wadi Halfa Area, *Nature, Lond.,* **203** : 341–343.
17. Sudan Museum (Sudan Government Antiquities Service), P.O. Box 178, Khartoum.
18. University of Colorado Museum, Boulder, Colorado, U.S.A.

1. Wadi Halfa (6B36).
2. 6B36 – cemetery site, 450 × 300 m, 152–155 m above sea level, 2·75 km N of Wadi Halfa, about 2·5 km from left bank of Nile.   21° 57′ N, 31° 19′ E.
3. G. W. Hewes, Director, H. T. Irwin and other members of the University of Colorado Nubian Expedition, 1962–1964.
4. Wind eroded desert surface, and in layer of *terra rossa* up to 30 cm thick.   G. W. Hewes, H. T. Irwin, M. Papworth and A. Saxe 1964, A new fossil human population from the Wadi Halfa area, Sudan, *Nature, Lond.,* **203** : 341–343.
5. ————
6. Late Pleistocene.   H. T. Irwin, J. B. Wheat and L. F. Irwin 1968, University of Colorado Investigations of paleolithic and epi-paleolithic sites in the Sudan, Africa, *Uni. Utah Anthrop. Pap,* No. 90.
7. Microlithic similar to Wadi complex with grinding stones.   H. T. Irwin *et al.* 1968. Qadan Industry.   F. Wendorf, J. L. Shiner and A. E. Marks 1965, Summary of the 1963–1964 field season, *Contributions to the Prehistory of Nubia,* Southern Methodist University Press : 9–35.
8. Bovids, fish and shellfish.   D. L. Greene and G. J. Armelagos 1971, *The Wadi Halfa Mesolithic Population,* Research Report No. 11, Dept. of Anthropology, University of Massachusetts.
9. ————
10. A3: 11950–6400 BP on basis of C14 dating of Qadan industry.   F. Wendorf, J. L. Shiner and A. E. Marks 1965.
11. **Wadi Halfa 6B36.**   **1–37** : 16 crania (i), 10 crania (ff), 82 dentes (i).
12. G. W. Hewes *et al.* 1964.
13. D. L. Greene, G. H. Ewing and G. J. Armelagos 1967, Dentition of a Mesolithic population from Wadi Halfa, Sudan, *Am. J. phys. Anthrop.,* **27** : 41–56.
14. D. L. Greene and G. J. Armelagos 1972.   Cf. Sahaba, Afalou, Taforalt.
15. D. L. Greene and G. J. Armelagos 1972.
16. G. J. Armelagos 1969, Disease in Ancient Nubia, *Science, N.Y.,* **163** : 255–259.   A. A. Saxe 1971, Social dimensions of mortuary practices in a mesolithic population from Wadi Halfa, Sudan, *Mem. Soc. Am. Archaeol.,* **25** : 39–57.
17. University of Colorado Museum, Boulder, Colorado, 80302, U.S.A.
18. University of Colorado Museum, Boulder, Colorado, 80302, U.S.A.

TANZANIA

Diana Margaret LEAKEY,

South African Museum,
P.O. Box 61,
Cape Town,
South Africa.

&

Mary Douglas LEAKEY,

Box 30239,
Nairobi,
Kenya.

EYASI

1. Eyasi.
2. NE shore of Lake Eyasi, about 2 km ($1\frac{1}{4}$ miles) on 11° bearing from W base of S peak of Mumba Hills. 3° 32′ S, 35° 16′ E.
3. L. Kohl-Larsen, 29 November 1935 and 1938.
4. Reddish, ill-consolidated estuarine sandstone, under 1·5–3·7 cm (4–12 inches) of surface, wind-borne sand. W. H. Reeve 1946, Geological Report on the Site of Dr Kohl-Larsen's Discovery of a Fossil Human Skull, Lake Eyasi, Tanganyika Territory, pt 2, *Jl E. Africa nat. Hist. Soc.* **19** : 44–50.
5. ————
6. Upper Pleistocene, Gamblian. W. H. Reeve 1946.
7. 'Levalloisian'. L. S. B. Leakey 1936, A new fossil skull from Eyasi, East Africa, *Nature, Lond.,* **138** : 1083. L. S. B. Leakey 1946, Report on a Visit to the Site of the Eyasi Skull, found by Dr Kohl-Larsen, pt 1, *Jl E. Africa nat. Hist. Soc.,* **19** : 40–43.
8. *Homoioceras, Bubalus* cf. *nilssoni.* L. S. B. Leakey 1946 : 41.
9. Eyasi 1: F = 1·30–1·53%, eU308 = 227–250 ppm, N = 0·01–0·02%. Eyasi 2: F = 2·0%, 100F/P205 = 8·9, eU308 = 208 ppm, N = 0·2%. Eyasi 3: F = 0·27–0·43%, eU308 = 58–71 ppm, N = 0·13–0·18%.
10. A1: *c.* 34,000 BP (1-UCLA-LJ) on basis of racemization of iso-leucine in hominid bone.
11. **Eyasi 1.** female?: cranium (ff), part of lt parietale, occipitale, lt temporale, part of frontale, maxilla (ff) with lt 2I alveoli, C, P3; loose upper rt M3. **Holotype** of *Palaeoanthropus njarasensis* Reck & Kohl-Larsen, 1936. Plate 15.
    Eyasi 2. male?: occipitale (f).
    Eyasi 3. occipitale (f, f).
12. L. Kohl-Larsen and H. Reck 1936, Erster Ueberblick über die Jungdiluvialen Tier und Menschenfunde Dr Kohl-Larsen's im Nordöstlichen Teil des Njarasa-Grabens (Ostafrika), *Geol. Rdsch.,* **27** : 401–441.
13. H. Weinert 1939, *Africanthropus njarasensis,* Beschreibung und phyletische Einordung des ersten Affenmenschen aus Ostafrika, *Z. Morph. Anthrop.,* **38** : 252–307.
14. R. Protsch 1977a, New morphological analysis based on a reconstruction and dating of the Eyasi hominids, *in* H. Müller-Beck (ed.) 1977, *Kohl-Larsen Festschrift,* Vol. 2.
15. H. Weinert 1939. R. Protsch 1977a.
16. L. H. Wells 1957, The Place of the Broken Hill Skull among Human Types, *Proc. 3rd Pan-Afr. Congr. Prehist.* : 173. P. V. Tobias 1968, Middle and Early Upper Pleistocene Members of the Genus *Homo* in Africa, *in* G. Kurth (ed.) 1968, *Evolution und Hominisation,* Stuttgart : 176–194. R. Protsch 1977b, The Kohl-Larsen Eyasi and Garusi Hominid Finds in Tanzania and Their Relation to *Homo erectus, in* B. A. Sigmon and J. S. Cybulski (eds) 1977, *Proc. Symp. on Homo erectus in honour of Davidson Black,* Toronto.
17. Department of Paleoanthropology and Archaeometry, FB–16, J. W. Goethe-Universität, 6000 Frankfurt am Main, Siesmayerstrasse 70, Germany (Eyasi 1, Eyasi 3). National Museum, P.O. Box 511, Dar es Salaam. (Eyasi 2: Reg. No. AH/65·2).
18. Anthropologisches Institut, Christian-Albrechts-Universität, Olshausenstrasse 40/60, Kiel, Federal Republic of Germany (Eyasi 1). British Museum (Natural History),

Cromwell Road, London SW7 5BD, England (Eyasi 2). Department of Palaeoanthropology and Archaeometry, FB–16, J. W. Goethe-Universität, 6000 Frankfurt am Main, Siesmayerstrasse 70, Germany.

GARUSI*

1. Garusi.
2. NW of Lake Eyasi, on the plateau near headwaters of Garusi River, Laetolil, 40 km N of site of Eyasi 1 and 2.   Approximately 3° 12′ S, 35° 11′ E.
3. L. Kohl-Larsen, 1939 (1–3).
4. On surface of subaerial tuffs of Laetolil (Vogel River) Beds.   P. E. Kent 1941, The Recent History and the Pleistocene Deposits of the Plateau North of Lake Eyasi, Tanganyika, *Geol. Mag.,* **78** : 173–184.
5. ————
6. Equivalent to basal Olduvai Bed II.   L. S. B. Leakey 1965, *Olduvai Gorge 1951–1961,* Cambridge, Vol. 1.   R. L. Hay 1976, *Geology of the Olduvai Gorge* : 187, Berkeley.   Stratigraphic horizon uncertain (Ed.).
7. No culture associated with hominid finds.
8. Fauna includes *Okapia stillei, Archidiskodon exoptatus.*   W. O. Dietrich 1941, Die säugetierpaläontologische Ergebnisse der Kohl-Larsen'schen Expedition, 1937–39 im nördlichen Deutsch-Ostafrika, *Cbl. Min. Geol. Paläont.,* **1941 B** : 217–223.   W. O. Dietrich 1950, Fossile Antilopen und Rinder Aquatorialafrikas, *Paläontographica,* **99,** A : 48–50.   Two faunal horizons are represented in the Laetolil area.   V. J. Maglio 1969, The status of the East African Elephant '*Archidiscodon exoptatus*' Dietrich 1942, *Brevoria,* no. 336 : 1·25.
9. Garusi 1 : F = 2·8%, eU308 = 308 ppm.   Garusi 2 : F = 2·7%, eU308 = 370 ppm, N = nil.
10. ————
11. **Garusi 1.**   rt maxilla (ff) with P3 and P4.   **Holotype** of *Meganthropus africanus* Weinert, 1950.   Plate 16.
    Garusi 2.   upper rt M3.
    Garusi 3.   rt maxilla (ff).
12. L. Kohl-Larsen 1943, *Auf den Spuren des Vormenschen,* Stuttgart, **2** : 379–381.   (Garusi 1–3).
13. H. Weinert 1950, Ueber die neuen Vor- und Frühmenschenfunde aus Afrika, Java, China und Frankreich, *Z. Morph. Anthrop.,* **42** : 138–142 (Garusi 1).   *Meganthropus africanus,* (Garusi 2 and 3).   *Africanthropus.*   A. Remane 1950, Die Zähne des *Meganthropus africanus, Z. Morph. Anthrop.,* **42** : 311–329.
14. R. Protsch 1977, A new morphological analysis based on a reconstruction and dating of the Garusi hominid, *in* H. Müller-Beck (ed.) 1977, *Kohl-Larsen Festschrift,* Vol. 2, *Homo* sp.
15. H. Weinert 1950.   R. Protsch 1977.
16. H. Weinert 1951, Ueber die Vielgestaltigkeit der Summoprimaten vor der Menschwer-

See also **LAETOLIL**

dung, *Z. Morph. Anthrop.*, **43** : 73–103, 311–330. J. T. Robinson 1953, *Meganthropus*, australopithecines and hominids, *Am. J. phys. Anthrop.*, **11** : 1–38 (Garusi 1 and 2). A. Remane 1954, Structure and Relationships of *Meganthropus africanus*, *Am. J. phys. Anthrop.*, **12** : 123–126. J. T. Robinson 1955, Further remarks on the relationship between '*Meganthropus*' and australopithecines, *Am. J. phys. Anthrop.*, **13** : 429–445. *Australopithecus africanus*.

17. Department of Palaeoanthropology and Archaeometry, FB–16, J. W. Goethe-Universität, 6000 Frankfurt am Main, Siesmayerstrasse 70, Germany.

18. Department of Palaeoanthropology and Archaeometry, FB–16, J. W. Goethe-Universität, 6000 Frankfurt am Main, Siesmayerstrasse 70, Germany. Sub-Department of Anthropology, British Museum (Natural History), Cromwell Road, London SW7 5BD, England (Garusi I).

LAETOLIL

1. Laetolil.
2. Open sites, faunal localities 1 (LH1), 3 (LH2), 5 (LH7, 12), 7 (LH3, 4, 6), 8 (LH5, 13), 1OW (LH10, 11), 11 (LH8), 19 (LH14), within an area of about 30 km², in the southern Serengeti Plains, at the northern margin of the Eyasi Plateau, in the divide between the Olduvai and Eyasi drainage systems.   3° 07′ – 3° 15′ S, 35° 10′ – 35° 15′ E.
3. M. Muoka, 1974 (LH1); M. Muluila, 1974 (LH2); M. Muoka, 1975 (LH3, 6, 7); M. Muluila, 1975 (LH4, 5); E. Kandini, 1975 (LH8, 10, 11, 12, 14); M. Jackes, 1975 (LH13).
4. Laetolil Beds.   P. E. Kent 1941, The recent history and the Pleistocene deposits of the plateau north of Lake Eyasi, Tankanyika, *Geol. Mag.*, **78** : 173–184.   Aeolian tuffs: in upper 30 m beneath a possible discontinuity.   R. L. Hay 1976, *Geology of the Olduvai Gorge*, Menlo Park: 187.   Laetolil Beds: upper part of aeolian tuffs underlying pale yellow vitric tuff (tuff *d*).   Below tuff *c* (LH1, 2, 3, 4, 6); above tuff *a* (LH8, 14); below tuff *a* (LH10, 11).   M. D. Leakey, R. L. Hay, G. H. Curtis, R. E. Drake, M. K. Jackes and T. D. White 1976, Fossil hominids from the Laetolil Beds, *Nature, Lond.*, **262** : 460–466.
5. ———
6. Upper Pliocene.
7. No trace of stone tools or even of utilised bone or stone.   M. D. Leakey *et al.* 1976.
8. Ceropithecinae; Colobinae; Viverridae; Hyaenidae; Felidae; machairodont; Rodentia: *Pedetes, Saccostomus, Hystrix;* Perissodactyla: *Hipparion, Ceratotherium, Diceros;* Artiodactyla: *Potamachoerus, Notochoerus, Ancylotherium, Orycteropus, Sivatherium, Giraffa jumae, Madoqua;* Proboscidea: *Deinotherium, Loxodonta;* Aves; Reptilia.   M. D. Leakey *et al.* 1976.   Two faunal horizons are represented in the Laetolil area.   V. J. Maglio 1969, The status of the East African Elephant '*Archidiscodon exoptatus*' Dietrich 1942, *Breviora*, no. 336: 1·25.
9. ———
10. A3: > 3·0 Myr on basis of K/Ar dating of overlying vogosite lava: 2·39 ± 0·09 Myr.   R. L. Hay 1976.   A3: > 3·59 Myr on basis of K/Ar and 40Ar/39Ar dating of tuff *c*. 3·41 ± 0·08, 3·68 ± 0·09, 3·69 ± 0·10 Myr (LH1, 2, 3, 4, 6, 14); < 3·77 Myr on

basis of K/Ar dating of biotite from tuff 50 m. below tuff *c.* : $3\cdot71 \pm 0\cdot12$, $3\cdot82 \pm 0\cdot16$ Myr (LH10, 11).   M. D. Leakey *et al.* 1976.

11. **LH 1.**   rt P4 (f).
    LH 2.   juvenile: corpus mandibulae, rt I1, I2, dc, dm1, dm2, M1; lt I1, I2, dc, dm1, dm2, M1.
    LH 3.   isolated dentes including upper dc, dm1, dm2, C.   Associated with LH6.
    LH 4.   adult: corpus mandibulae, rt C–M3; lt C (f), P3 (f), P4–M2.
    LH 5.   adult: upper I2–M1.
    LH 6.   juvenile: isolated upper permanent and deciduous dentes.   Associated with LH3.
    LH 7.   rt M1 or 2 (f).
    LH 8.   rt M2, M3.
    LH10.   lt corpus mandibulae (f), dentes (roots).
    LH11.   lt M1 or 2.
    LH12.   lt M1 or 2 (f).
    LH13.   rt corpus mandibulae (f), dentes (roots).
    LH14.   isolated lower dentes.
12. M. D. Leakey *et al.* 1976.
13. M. D. Leakey *et al.* 1976.   Aff. *Homo.*
14. T. D. White 1977, New Fossil Hominids from Laetolil, Tanzania, *Am. J. phys. Anthrop.,* **46** : 197–216.
15. M. D. Leakey *et al.* 1976.   (LH2, 4).   T. D. White 1977.
16. —————
17. National Museum, P.O. Box 511, Dar es Salaam.
18. —————

NATRON

See under **PENINJ**

NDUTU

1. Ndutu.
2. Open site on N shore of Lake Ndutu, Serengeti Plains.   3° 00′ S, 35° 00′ E.
3. A. A. Mturi and Tanzanian Department of Antiquities, September–October 1973.
4. 1st occupation floor on silty clay sub-unit of greenish sandy clay unit of Ndutulu Lacustrine deposits.   A. A. Mturi 1976, New hominid from Lake Ndutu, Tanzania, *Nature Lond.,* **262** : 484–485.
5. —————
6. Middle Pleistocene, cf. Masek Beds of Olduvai Gorge.   A. A. Mturi 1976.
7. Spheroids, hammerstones, flakes etc., associated with possible butchery site.   A. A. Mturi 1976.
8. Extensive faunal debris, undescribed.
9. —————
10. A2: 500,000–600,000 BP on basis of racemization measurement of associated bone. A. A. Mturi 1976.
    A3: 0·4–0·6 Myr on basis of K/Ar dating and palaeomagnetic polarity (normal) of

Masek Beds at Olduvai Gorge.   R. L. Hay 1976, *Geology of the Olduvai Gorge:* 136–137, Menlo Park.
11.   **Ndutu 1.**   cranium (f).
12.   A. A. Mturi 1976.
13.   R. J. Clark 1976, New cranium of *Homo erectus* from Lake Ndutu, Tanzania, *Nature, Lond.,* **262** : 485–487.   *Homo erectus.*
14.   ⸺⸺⸺
15.   R. J. Clark 1976.
16.   ⸺⸺⸺
17.   National Museum of Tanzania, P.O. Box 511, Dar es Salaam.
18.   ⸺⸺⸺

OLDUVAI GORGE

General Entry

1.   Olduvai Gorge.
2.   River Gorge in the Serengeti Plains at the western margin of the Eastern Rift Valley, NNW of Ngorongoro Crater, 160 km NW of Arusha, Northern Tanzania.   The distribution of sites is shown in Fig. 3.   They fall within the area 2° 56′ – 2° 59′ S, 35° 15′ – 35° 22′ E.
4.   Interbedded volcanic and sedimentary deposits of alluvial, lacustrine and aeolian origin subdivided into Beds I–IV, Masek, Ndutu and Naisiusiu.   R. L. Hay 1976, *Geology of the Olduvai Gorge,* Berkeley.
5.   ⸺⸺⸺
6.   Lower to uppermost Pleistocene.   R. L. Hay 1976.
7.   Oldowan and Acheulian cultures in Beds I–IV.   M. D. Leakey 1971a, *Olduvai Gorge,* Vol. 3, Cambridge: 9–16.   M. D. Leakey 1971b, Discovery of Postcranial Remains of *Homo erectus* and Associated Artefacts in Bed IV at Olduvai Gorge, Tanzania, *Nature, Lond.,* **232** : 380–383.   M. D. Leakey, R. L. Hay, D. L. Thurber, R. Protsch and R. Berger 1972, Stratigraphy, archaeology, and the age of the Ndutu and Naisiusiu Beds, Olduvai Gorge, Tanzania, *Wld Archaeol.,* **3** : 328–341.
8.   Extensive fauna including numerous species of Mammalia, Reptilia, Aves, Amphibia, Pisces and Mollusca.   L. S. B. Leakey 1965, *Olduvai Gorge 1951–61,* Vol. 1: 1–118.   A. W. and A. Gentry 1977, Fossil Bovidae (Mammalia) of Olduvai Gorge, Tanzania, *Bull. Br. Mus. nat. Hist.,* (Geol.) **30.**   P. M. Butler and M. Greenwood 1976, *in* R. J. G. Savage and S. C. Coryndon (eds) 1976, Elephant Shrews (Macroscelididae) for Olduvai and Makapansgat, *Fossil Vertebrates of Africa,* Vol. 4: 1–56.
9.   See under individual entries.
10.   Upper Bed I: 1·8–1·7 Myr on basis of 57 K/Ar dates of tuffs and pumice distributed throughout Bed.
Bed II: 1·7–1·15 Myr on basis of K/Ar date of Tuff IIa, palaeomagnetic data and relative stratal thickness.   Lemuta Member: *c.* 1·6 Myr.
Bed III/IV: 1·15–0·6 Myr on basis of palaeomagnetic data and relative stratal thickness.
Masek Beds: 0·6–0·4 Myr on basis of palaeomagnetic data and K/Ar dating.
Ndutu Beds: Lower Unit 400,000–60,000 BP
                Upper Unit 60,000–32,000 BP both on basis of C14, amino-acid racemiza-

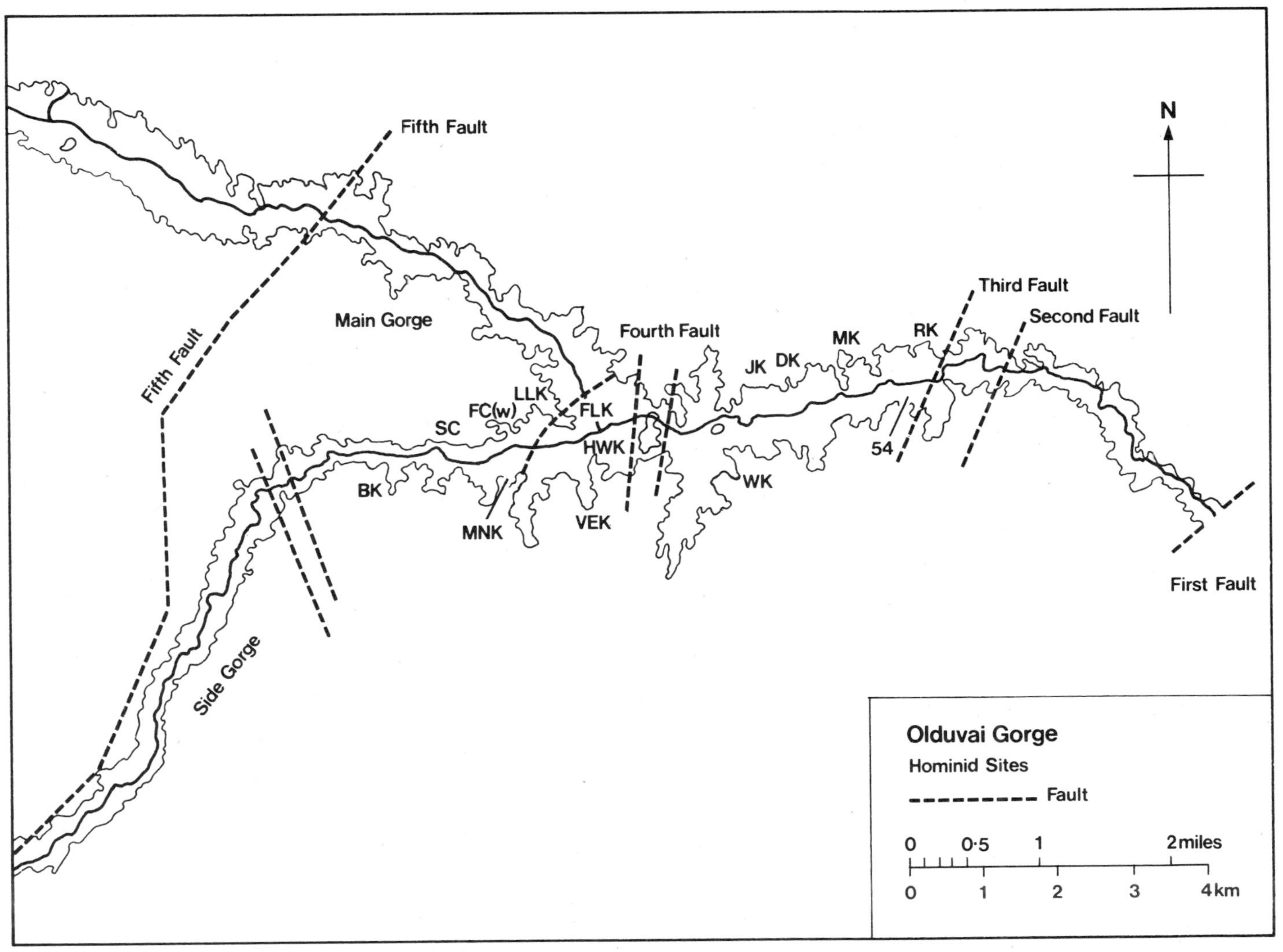

Fig. 3.   Plan of Olduvai Gorge showing hominid sites.

tion and geochemical data (e.g. alteration of napheline).

Naisiusiu Beds: > 22,000 − < 15,000 BP on basis of C14 dates obtained from calcretes and bone collagen.   R. L. Hay 1976.

The correlation of Magnetic epochs etc. is shown in Fig. 4.

11–15.   See under individual entries.

16.   H. Reck 1926, Prähistorische Grab- und Menschenfunde und ihre Beziehungen zur Pluvialzeit in Ostafrika, *Mitt. dt. Schutzgeb.,* **34** : 81–86.   L. S. B. Leakey 1931, *The Stone Age Cultures of Kenya Colony,* Cambridge.   L. S. B. Leakey 1945b, *The Stone Age Races of Kenya,* London.   L. S. B. Leakey 1951, *Olduvai Gorge: a Report on the Evolution of the Handaxe Cultures in Beds I–IV,* London.   L. S. B. Leakey 1953, *Adam's Ancestors,* 4th Edn., London.

17.   National Museum, P.O. Box 511, Dar es Salaam.

18.   While stocks last, casts of the most important specimens may be bought from the Wenner-Gren Foundation for Anthropological Research, 14 East 71st Street, New York, N.Y. 10021, U.S.A. (OH5, OH6, OH8, OH9, OH13, OH48).   Casts of all the specimens except OH1, OH2 may be bought from National Museum, P.O. Box 40658, Nairobi, Kenya.

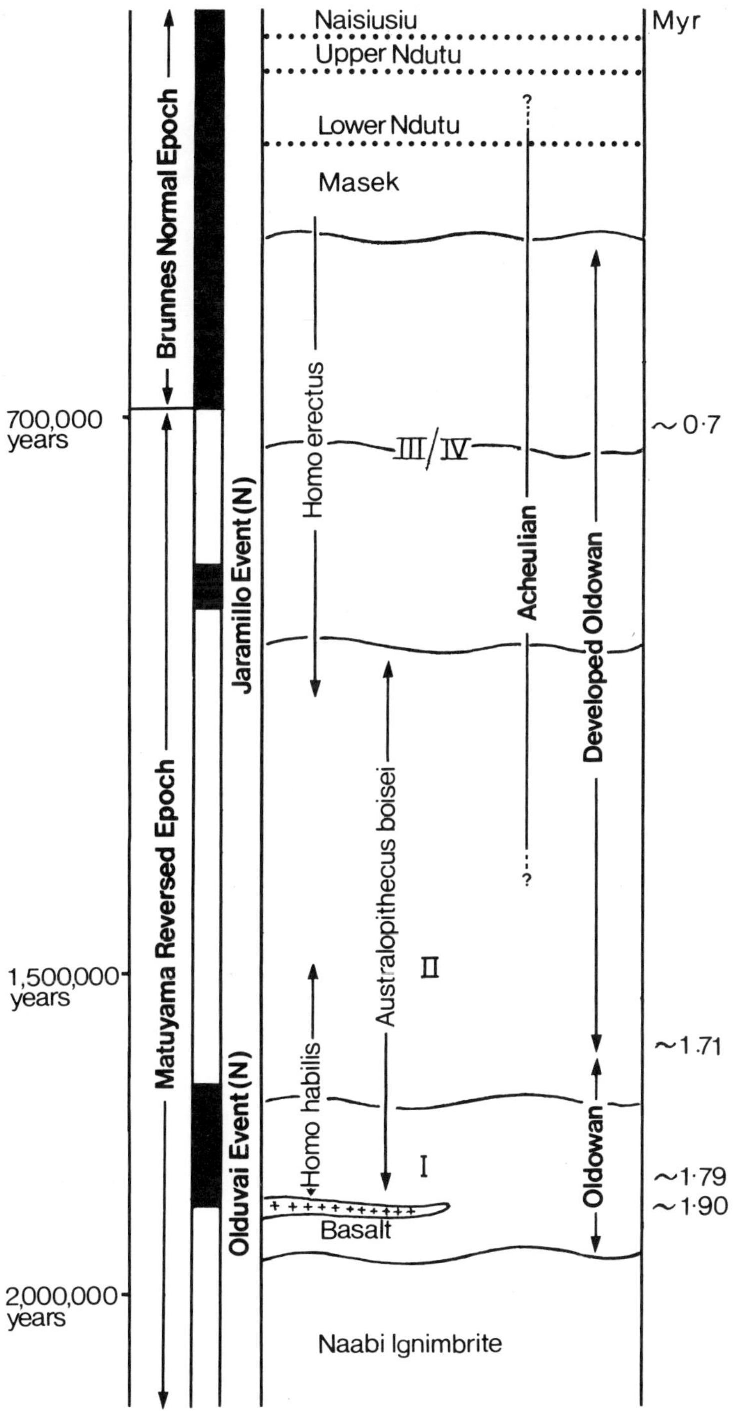

Fig. 4. Correlation of magnetic epochs, stratigraphic beds, cultures, and hominids at Olduvai Gorge.

OLDUVAI GORGE

Sites are given in alphabetical order under headings of the various Beds.   Beds themselves are in the order I, II, III/IV, Masek, Ndutu, Naisiusiu and Unknown.   The following list is intended to facilitate the location of individual hominids.

Reg. No.	*Bed and Site*	*Element*	*Page*
OH 1	Naisiusiu, site RK	skeleton	184
OH 2	Bed IV, site MNK	parietale	181
OH 3	Bed II, site BK	dentes	176
OH 4	Bed I, site MK	mandibula	175
OH 5	Bed I, site FLK	cranium	172
OH 6	Bed I, site FLK	parietale, dentes	172
OH 7	Bed I, site FLK NN	parietale, mandibula, ossa manus	173
OH 8	Bed I, site FLK NN	ossa pedis	173
OH 9	Bed II, site LLK	calvaria	179
OH 10	Bed I, site FLK North	os pedis	173
OH 11	Lower Ndutu Beds, site DK	palatinum, dentes	183
OH 12	Bed IV, site VEK	calotte, palatinum, dentes	182
OH 13	Bed II, site MNK	calotte, palatinum, mandibula	179
OH 14	Bed II, site MNK	parietale, frontale	179
OH 15	Bed II, site MNK	dentes	179
OH 16	Bed II, site FLK Maiko Gully	calotte, dentes	177
OH 17	Unknown, site FLK Maiko Gully	dens	185
OH 19	Bed II, site FC West	M	177
OH 20	Bed II, site HWK	femur	178
OH 21	Unknown, site FLK North	M	185
OH 22	Bed IV, unnamed site	mandibula, dentes	183
OH 23	Masek Beds, site FLK	mandibula	183
OH 24	Bed I, site DK East	cranium, dentes	171
OH 25	Unknown, geologic locality 54	parietale	185
OH 26	Bed II, site FLK West	M	178
OH 27	Bed I, site HWK EE	M	175
OH 28	Bed IV, site WK	femur, os coxae	182
OH 29	Bed III, site JK	M	181
OH 30	Bed II, site FLK Maiko Gully	cranial fragments, dentes	177
OH 31	Bed I, site HWK East	M	175
OH 32	Bed II, site MNK	dens	179
OH 33	Bed I, site FLK NN	parietale, occipitale	173
OH 34	Bed III, site JK	femur, tibia	181
OH 35	Bed I, site FLK	tibia, fibula	172
OH 36	Bed II, site SC	ulna	180
OH 37	Bed II, site FLK	mandibula	177
OH 38	Bed II, site SC	2 dentes	180
OH 39	Bed I, site HWK EE	dentes	175
OH 40	Bed II, site FLK	dens	177
OH 41	Bed II, site HWK EE	dens	179
OH 42	Bed I, site HWK EE	dens	175
OH 43	Bed I, site FLK NN	2 metatarsalis	173
OH 44	Bed I, site FLK	M	172

Reg. No.	Bed and Site	Element	Page
OH 45	Bed I, site FLK NN	M	173
OH 46	Bed I, site FLK NN	P4	173
OH 48	Bed I, site FLK NN	clavicula	173
OH 49	Bed I, site FLK NN	radius	173
OH 50	Bed I, site FLK NN	costa	173
OH 51	Bed I, site FLK NN	ossa manus	173
OH 52	Bed I, site FLK NN	os manus	173

OH 18, OH 47 are not hominid.

BED I

1. DK East.
2. Erosion slope, N side of Main Gorge, about 3 km E of junction with Side Gorge.   2° 59′ S, 35° 22′ E.
3. P. Nzube, assistant to M. D. Leakey, 1968 (OH24).
4. Surface of Lower Bed I, inferred *ex* tuffaceous clay underlying Tuff IB (OH24).   M. D. Leakey, R. J. Clarke and L. S. B. Leakey 1971, New hominid skull from Bed I, Olduvai Gorge, Tanzania, *Nature, Lond.*, **232** : 308–312.
6. Lower Pleistocene.
7. Inferred Oldowan.   M. D. Leakey 1971, *Olduvai Gorge,* Vol. 3, Cambridge.
8. *Crocodylus, Elephas recki* stage 3, *Deinotherium* cf *bozasi, Parmularius altidens.*   L. S. B. Leakey 1965.
10. A3: *c.* 1·8 Myr (KA 1057) on basis of K/Ar dating of minerals in overlying Tuff IB. G. H. Curtis and R. L. Hay 1972, Further geological studies and potassium-argon dating at Olduvai Gorge and Ngorongoro Crater, *in* W. W. Bishop and J. A. Miller (eds) 1972, *Calibration of Hominoid Evolution,* Edinburgh and Toronto : 289–301.
11. **OH24.**   cranium (ff), lt P3 (f), P4, M1, roots of I1, I2; roots of rt I1, I2, rt M1–3 (all f), dentes (ff).
12. M. D. Leakey 1969, Recent discoveries of hominid remains at Olduvai Gorge, Tanzania, *Nature, Lond.*, **223** : 756.
13. M. D. Leakey, R. J. Clarke and L. S. B. Leakey 1971 (OH24).   *Homo.*   P. V. Tobias, in preparation.
14. P. V. Tobias, in preparation (OH24).
15. M. D. Leakey, R. J. Clarke and L. S. B. Leakey 1971 (OH24).
16. P. V. Tobias 1972, ' Dished faces ', brain size and early hominids, *Nature, Lond.*, **239** : 468–469.   R. L. Holloway 1973, New endocranial values for Olduvai and East Rudolf hominids, *Nature, Lond.*, **243** : 97–99.   R. L. Holloway 1973, Endocranial Volumes of Early African Hominids, and the role of the Brain in Human Mosaic Evolution, *J. hum. Evol.*, **2** : 449–459.

1. FLK.
2. Occupation floor and erosion slope below, on S side of Main Gorge just W of junction with Side Gorge.   2° 59′ S, 35° 20′ E.
3. M. D. Leakey, 17 July 1959 (OH 5); assistants to L. S. B. Leakey, 1960 (OH 6, OH35); E. Kandini, assistant to M. D. Leakey, January 1970 (OH44).
4. *In situ* on clay palaeosol between Tuffs IB and ID, 5·8 m (19 ft) above basalt and 6·7 m (22 ft) below top of Bed I (OH 5, OH35); surface, presumed same horizon as OH 5 and H.35 (OH 6, OH44).   M. D. Leakey 1971, *Olduvai Gorge,* Vol. 3, Cambridge.
6. Lower Pleistocene.
7. Oldowan.   M. D. Leakey 1966, A review of the Oldowan culture from Olduvai Gorge, Tanzania, *Nature, Lond.,* **210** : 462–466.   M. D. Leakey 1971.
8. *Galago senegalensis, Crocuta* aff. *ultra, Promesochoerus mukiri, Kobus sigmoidalis, Parmularius altidens, Gazella wellsi.*   M. Leakey 1971 *in* M. D. Leakey 1971.
9. OH35 tibia: F = 2·5%, 100F/P205 = 9·8, eU308 = 340 ppm, N = Nil.
10. A3: between 1·68 (KA-851) and 2·04 Myr (KA-1036) on the basis of K/Ar dating of minerals in tuffs between Tuffs IB and ID.   G. H. Curtis and R. L. Hay 1972, Further geological studies and potassium-argon dating at Olduvai Gorge and Ngorongoro Crater, *in* W. W. Bishop and J. A. Miller (eds) 1972, *Calibration of Hominoid Evolution,* Edinburgh and Toronto : 289–301.
11. **OH 5.**   young male: cranium with complete dentition.   **Holotype** of *Zinjanthropus boisei* Leakey, 1959.   Plate 17.   Reg. No. AH/65.1.
    OH 6.   parietale (f), upper rt M1, upper rt I2, lower lt P3.
    OH 35.   lt tibia (f), lt fibula (f).
    OH 44.   crown of upper rt M1.
12. L. S. B. Leakey 1959, A new fossil skull from Olduvai, *Nature, Lond.,* **184** : 491–493.   L. S. B. Leakey 1960a, Recent discoveries at Olduvai Gorge, *Nature, Lond.,* **188** : 1050–1052 (OH 6 & OH35).
13. P. V. Tobias 1967, *Olduvai Gorge,* Vol. 2, Cambridge (OH 5).   *Australopithecus (Zinjanthropus) boisei.*   L. S. B. Leakey, P. V. Tobias and J. R. Napier 1964, A new species of the genus *Homo* from Olduvai Gorge, *Nature, Lond.,* **202** : 7–9 (OH 6).   *Homo habilis* paratype.   P. R. Davis, M. H. Day and J. R. Napier 1964, Hominid fossils from Bed I, Olduvai Gorge, Tanganyika, *Nature, Lond.,* **201** : 967–970 (OH35).   *Homo habilis.*
14. ————
15. P. V. Tobias 1967 (OH 5).   P. R. Davis, M. H. Day and J. R. Napier 1964 (OH35).
16. J. T. Robinson 1960, The affinities of the new Olduvai australopithecine, *Nature, Lond.,* **186** : 456–458.   L. S. B. Leakey 1960, The affinities of the new Olduvai australopithecine (Reply to J. T. Robinson), *Nature, Lond.,* **186** : 458.   G. Kurth 1960, *Zinjanthropus boisei* aus dem Unterpleistozan von Oldoway, Ostafrika, *Sonderh. Naturw.,* **12** : 265.   P. V. Tobias 1963, Cranial capacity of *Zinjanthropus* and other australopithecines, *Nature, Lond.,* **197** : 743–746.   J. T. Robinson 1965, *Homo 'habilis'* and the australopithecines, *Nature, Lond.,* **205** : 121–124.   P. V. Tobias 1965a, Early man in East Africa, *Science, N.Y.,* **149** : 22–33 (OH 5, OH 6).   P. V. Tobias 1965b, New discoveries in Tanganyika, their bearing on hominid evolution, *Curr. Anthrop.,* **6** : 391–411 (OH 5).   H. J. Pichler 1971, Zur Anatomie des Schädels von Australopithecus (Zinjanthropus) boisei aus der Olduvai schlucht in Ostafrica, *Archiv. klin. exp. Ohren-, Nasen- kehlkopfheilk.,* **199** : 649–659.   R. L. Holloway 1973, Endocranial Volumes of Early African Hominids, and the Role of the Brain in Human Mosaic Evolution, *J. hum. Evol.,* **2** : 449–459.

1. FLK North.
2. Living floor at site FLK North, about 180 m (200 yards) N of site FLK.   2° 59′ S, 35° 20′ E.
3. J. Mutaba, assistant to L. S. B. Leakey, 1961.
4. *In situ* in clays 1·5 m (5 ft) below Tuff IF at top of Bed I.   M. D. Leakey 1971, *Olduvai Gorge*, Vol. 3, Cambridge.
6. Lower Pleistocene.
7. Oldowan.   M. D. Leakey 1971.
8. *Deinotherium* cf. *bozasi, Elephas recki* stage 3, *Ectopotamochoerus dubius, Parmularius altidens, Gorgon olduvaiensis.*   M. Leakey 1971, *in* M. D. Leakey 1971.
10. ─────────
11. **OH10.**   rt phalanx distalis pedis I.
12. P. V. Tobias 1965a, Early Man in East Africa, *Science N.Y.,* **149** : 22–33.
13. M. H. Day and J. R. Napier 1966, A hominid toe bone from Olduvai Gorge, Tanzania, *Nature, Lond.,* **211** : 929–930.   Cf. *Homo habilis.*
14. ─────────
15. ─────────
16. M. H. Day 1967, Olduvai Hominid 10: a multivariate analysis, *Nature, Lond.,* **215** : 323–324.   M. H. Day 1974, The Interpolation of isolated fossil Foot Bones into a Discriminant Function Analysis—A Reply, *Am. J. phys. Anthrop.,* **41** : 233–236.   B. A. Wood 1974, Olduvai Bed I Post-cranial Fossils: A Reassessment, *J. hum. Evol.,* **3** : 373–378.

1. FLK NN.
2. Erosion slope and on occupation floor at site FLK NN, about 365 m (400 yds) N of site FLK.   2° 59′ S, 35° 20′ E.
3. J. H. E. Leakey, 2 November 1960 (OH 7); P. Nzube, assistant to M. D. Leakey, 28 November 1969 (OH33); assistants to L. S. B. Leakey, 1960 (OH 8, 43, 45, 46, 48–52).
4. *In situ* on palaeosol in Bed I about 4·5 m (15 ft) above basalt and slightly below level of OH 5 at site FLK (OH 7, 8, 43, 48–52); surface, presumed same horizon (OH33, 45, 46).   M. D. Leakey 1971, *Olduvai Gorge,* Vol. 3, Cambridge.
5. ─────────
6. Lower Pleistocene.
7. Oldowan.   M. D. Leakey 1971.
8. *Simopithecus* cf. *oswaldi, Otocyon recki, Pseudocivettictis ingens, Potamochoerus intermedius, Gazella wellsi.*   M. Leakey 1971, *in* M. D. Leakey 1971.
9. OH 8 tarsus: eU308 = 150 ppm.
10. A3: *c.* 1·78 Myr (KA-850) on basis of K/Ar dating of ash 2·5 cm above occupation floor.   L. S. B. Leakey, J. F. Evernden and G. H. Curtis 1961, Age of Bed I, Olduvai Gorge, Tanganyika, *Nature, Lond.,* **191** : 478–479.   G. H. Curtis and R. L. Hay 1972, Further geological studies and potassium-argon dating at Olduvai Gorge and Ngorongoro Crater, *in* W. W. Bishop and J. A. Miller (eds) 1972, *Calibration of Hominoid Evolution,* Edinburgh and Toronto : 289–301.
11. **OH 7.**   juvenile: lt and rt parietalia (ff); mandibula, lt I1–M2, rt I1–M1; phalanges proximalis II and IV, phalanges media II–V, phalanx distalis I and 2

others. **Holotype** of *Homo habilis* Leakey, Tobias and Napier, 1964. Plate 18.

Associated bones (probably the same individual): rt scaphoideum, rt trapezium, lt capitatum, metacarpale II (f), fragments. (Associated proximal phalanges of an adult are probably not hominid).

OH 8. lt pes comprising 7 ossa tarsi and 5 metatarsalia.

OH 33. 5 fragments of parietale and occipitale.

OH 43. (originally published as H. 8). lt metatarsalia III, V.

OH 45. (originally published as H. 7). upper lt M1.

OH 46. (originally published as H. 8). probably distal part crown upper rt P4.

OH 48. (originally published as H. 8). lt clavicula.

OH 49. (originally published as H. 8). diaphysis radius (f).

OH 50. (originally published as H. 8). costa (f).

OH 51. (originally published as H. 8). rt scaphoideum, trapezium, base metacarpalis II.

OH 52. (originally published as H. 8). lt capitatum.

12. L. S. B. Leakey 1960, Recent discoveries at Olduvai Gorge, *Nature, Lond.*, **188** : 1050 (OH 8, 43, 48–52). L. S. B. Leakey 1961a, New Finds at Olduvai Gorge, *Nature, Lond.*, **189** : 649–650. M. D. Leakey 1971 (OH33).

13. L. S. B. Leakey 1961b, The juvenile mandible from Olduvai, *Nature, Lond.*, **191** : 417–418 (OH 7 mandibula). J. R. Napier 1962, Fossil hand bones from Olduvai Gorge, *Nature, Lond.*, **196** : 409–411 (OH 7 ossa manus). P. R. Davis, M. H. Day and J. R. Napier 1964, Hominid fossils from Bed I, Olduvai Gorge, Tanganyika, *Nature, Lond.*, **201** : 967–970 (OH 8, 43).

14. B. A. Wood 1974, Olduvai Bed I Post-cranial Fossils: A Reassessment, *J. hum. Evol.*, **3** : 373–378.

15. L. S. B. Leakey 1960c. L. S. B. Leakey 1961a. P. R. Davis, M. H. Day and J. R. Napier 1964. P. V. Tobias 1964, The Olduvai Bed I hominine with special reference to its cranial capacity, *Nature, Lond.*, **202** : 3–4.

16. L. S. B. Leakey, P. V. Tobias and J. R. Napier 1964, A new species of the genus *Homo* from Olduvai Gorge, *Nature, Lond.*, **202** : 7–9 (OH 7, 8). *Homo habilis.* W. E. Le Gros Clark 1964, The Evolution of Man, letter in *Discovery, Lond.*, **25** : 49. P. V. Tobias and G. H. R. von Koenigswald 1964, A comparison between the Olduvai hominines and those of Java and some implications for hominid phylogeny, *Nature, Lond.*, **204** : 515–518. P. V. Tobias 1965a, Early man in East Africa, *Science, N.Y.*, **149** : 22–33. P. V. Tobias 1965b, New discoveries in Tanganyika, their bearing on hominid evolution, *Curr. Anthrop.*, **6** : 391–411. R. L. Holloway Jr 1965, Cranial capacity of the hominine from Olduvai Bed I, *Nature, Lond.*, **208** : 205–206. R. L. Holloway Jr 1966, Cranial capacity of the Olduvai Bed I hominine, *Nature, Lond.*, **210** : 1108–1109. L. S. B. Leakey 1966, *Homo habilis, Homo erectus* and the australopithecines, *Nature, Lond.*, **209** : 1279–1281. P. V. Tobias 1966, The distinctiveness of *Homo habilis, Nature, Lond.*, **209** : 953–957. M. H. Day and B. A. Wood 1968, Functional affinities of the Olduvai Hominid 8 talus, *Man*, **3** : 440–455. R. L. Holloway 1973, Endocranial Volumes of Early African Hominids, and the Role of the Brain in Human Mosaic Evolution, *J. hum. Evol.*, **2** : 449–459.

1.   HWK East.
2.   Erosion slope on S side of Main Gorge just E of junction with Side Gorge.   2° 59′ S, 35° 20′ E.
3.   M. Mutala, assistant to M. D. Leakey, 19 October 1969.
4.   Surface, *ex* clay below Tuff IF at top of Bed I.
6.   Lower Pleistocene.
7.   Inferred Oldowan.   M. D. Leakey 1971, *Olduvai Gorge,* Vol 3, Cambridge.
11. **OH31.**   lower rt M1 or M2 (f).
12. M. D. Leakey 1971.
13. P. V. Tobias, in preparation.
14. ─────────
15. ─────────

1.   HWK EE.
2.   Erosion slope about 180 m (200 yds) E of site HWK East.   2° 59′ S, 35° 20′ E.
3.   B. Musembe, 1969 (OH27); N. Mbuika, 8 January 1972 (OH39); E. Kandini, 1972 (OH42), assistants to M. D. Leakey.
4.   Surface, Upper Bed I (OH27); surface, in algal concretion *ex* 0·6 m (2 ft) below Tuff IF at top of Bed I (OH39); surface, *ex* 1·25 m (4 ft) below Tuff IF (OH42).
6.   Lower Pleistocene.
7.   Inferred Oldowan.   M. D. Leakey 1971, *Olduvai Gorge,* Vol. 3, Cambridge.
11. **OH27.**   lower rt M3.
     OH39.   upper lt dc–dm2 (all f), I1–M2: upper rt dm1, dm2, C (crown), M2.
     OH42.   upper rt P3 or P4 (f).
12. M. D. Leakey 1969, Recent discoveries of hominid remains at Olduvai Gorge, Tanzania, *Nature, Lond.,* **223** : 756.
13. P. V. Tobias, in preparation.
14. ─────────
15. ─────────

1.   MK.
2.   Erosion slope on N side of Main Gorge about 4 km E of junction with Side Gorge.   2° 59′ S, 35° 22′ E.
3.   H. Mukiri, assistant to L. S. B. Leakey, 1959.
4.   Surface and *in situ* in clays beneath Tuff IB, 2·7 m (9 ft) above basalt of Bed I.   M. D. Leakey 1971.
6.   Lower Pleistocene.
7.   Oldowan.   M. D. Leakey 1971, *Olduvai Gorge,* Vol. 3, Cambridge.
8.   *Crocodylus, Lutra, Elephas recki* stage 3.   M. Leakey 1971, *in* M. D. Leakey 1971.
10. A3: between 1·75 and 1·85 Myr. on basis of K/Ar dating of samples low in Bed I.   J. F. Evernden and G. H. Curtis 1965, Potassium-argon dating of late Cenozoic rocks in East Africa and Italy, *Curr. Anthrop.,* **6** : 343–385.   G. H. Curtis and R. L. Hay 1972,

Further geological studies and potassium-argon dating at Olduvai Gorge and Ngorongoro Crater, *in* W. W. Bishop and J. A. Miller (eds) 1972, *Calibration of Hominoid Evolution,* Edinburgh and Toronto : 289–301.

11. **OH 4.** lt corpus mandibulae (f), M3 (or M2), isolated lower P and M (f).
12. L. S. B. Leakey, J. F. Evernden and G. H. Curtis 1961, Age of Bed I, Olduvai, *Nature, Lond.,* **191** : 478–479.
13. L. S. B. Leakey, P. V. Tobias and J. R. Napier 1964, A new species of the genus *Homo* from Olduvai Gorge, *Nature, Lond.,* **202** : 7–9. *Homo habilis.*
14. —————
15. —————
16. J. T. Robinson 1965, *Homo 'habilis'* and the australopithecines, *Nature, Lond.,* **205** : 121–124. P. V. Tobias 1966, The distinctiveness of *Homo habilis, Nature, Lond.,* **209** : 953–957. J. T. Robinson 1966, The distinctiveness of *Homo habilis* (Reply to P. V. Tobias), *Nature, Lond.,* **209** : 957–960.

BED II

1. BK.
2. Excavation on S side of Side Gorge about 3 km W, of junction with Main Gorge. 2° 59′ S, 35° 19′ E.
3. An assistant to L. S. B. Leakey, 1955.
4. *In situ* in Upper Bed II. L. S. B. Leakey 1964, The new Olduvai hominid discoveries and their geological and faunal siting, *Proc. geol. Soc. Lond.,* No. 1617 : 104–109. M. D. Leakey 1971, *Olduvai Gorge,* Vol. 3, Cambridge.
6. Lower Pleistocene.
7. Developed Oldowan B. M. D. Leakey 1971.
8. *Elephas recki* stage 3, *Mesochoerus olduvaiensis, Pelorovis oldowayensis, Libytherium olduvaiensis, Stylohipparion, Hippopotamus gorgops.* M. Leakey 1971, *in* M. D. Leakey 1971.
11. **OH 3.** dc, dm2.
12. L. S. B. Leakey 1958, Recent discoveries at Olduvai Gorge, Tanganyika, *Nature, Lond.,* **181** : 1099–1103.
13. P. V. Tobias, in preparation.
14. —————
15. L. S. B. Leakey 1958.
16. G. H. R. von Koenigswald 1960, Remarks on a fossil human molar from Olduvai, East Africa, *Proc. K. ned. Akad. Wet.,* **63B** : 20–25. A. A. Dahlberg 1960, The Olduvai giant hominid tooth, *Nature, Lond.,* **188** : 962. L. S. B. Leakey 1960, An alternative interpretation of the supposed giant deciduous hominid tooth from Olduvai (Reply to J. T. Robinson), *Nature, Lond.,* **185** : 408. J. T. Robinson 1960, An alternative interpretation of the supposed giant deciduous hominid tooth from Olduvai *Nature, Lond.,* **185** : 407–408.

1.   FC West.
2.   Excavation on N side of Side Gorge about 2 km W of junction with Main Gorge.
     2° 59' S, 35° 19' E.
3.   An assistant to L. S. B. Leakey, 1963.
4.   *In situ* in re-worked tuff 4·6 m (15 ft) above chert horizon in Upper Middle Bed II.   M.
     D. Leakey 1971, *Olduvai Gorge,* Vol. 3, Cambridge.
6.   Lower Pleistocene.
7.   Developed Oldowan.   M. D. Leakey 1971.
8.   *Clarias, Struthio, Stylohipparion, Hippopotamus gorgops, Libytherium olduvaiensis.*
     M. Leakey 1971, *in* M. D. Leakey 1971.
11. **OH19.**   M (f).
12. M. D. Leakey 1971.
13. P. V. Tobias, in preparation.
14. ————
15. ————

1.   FLK.
2.   On road up to 1971 camp (OH37) and on erosion slopes above site of OH 5 (see above,
     Bed I) and at S end of FLK exposures.   2° 59' S, 35° 20' E.
3.   M. D. Leakey, 1971 (OH37); M. Mwoka, assistant to M. D. Leakey, 1972 (OH40).
4.   Surface, probably *ex* Sandy Conglomerate in Lower Bed II (OH37, 40).
6.   Lower Pleistocene.
11. **OH37.**   lt corpus mandibulae, M1–2, roots I2, C, P4 and M3.
     OH40.   upper ?rt M1 (f).
12. ————
13. ————
14. ————
15. ————

1.   FLK, Maiko Gully.
2.   Erosion slope about 180 m (200 yds) S of site FLK.   2° 59' S, 35° 20' E.
3.   M. Matumbo, assistant to L. S. B. Leakey, 1963 (OH16); M. Mutala and E. Kandini,
     assistants to M. D. Leakey, 1969 (OH30).
4.   Surface, eroded from clays low in Bed II.   *In situ* at base of clay with roots cases,
     0·7 m (2 ft) above Tuff IF at top of Bed I (OH16).   Surface, Lower Bed II, inferred
     same horizon as OH16 (OH30).   L. S. B. Leakey and M. D. Leakey 1964, Recent dis-
     coveries of fossil hominids in Tanganyika: at Olduvai and near Lake Natron, *Nature,
     Lond.,* **202** : 5–7.   M. D. Leakey 1969, Recent discoveries of hominid remains at
     Olduvai Gorge, Tanzania, *Nature, Lond.,* **210** : 462 (OH16).

6.   Lower Pleistocene.
7.   Oldowan.   M. D. Leakey 1971, *Olduvai Gorge,* Vol. 3, Cambridge.
11.   **OH 16.**   young adult: calotte (ff); upper rt I1, I2, C (ff), P3–M3 and lt I1 (ff), I2 (ff),
      P3, P4, M1 (ff), M2, M3; lower rt I2, C (f), P3–M3 and lt I1, I2, P3 (ff), P4,
      M1 (f), M2 (f), M3.
      OH 30.   cranial fragments, upper rt C, M1 (f), dm1, dm2 (f) and lt C (f), P3, P4 (f),
      M1, dm1 (ff); lower rt I1, I2 and lt I1, C, M1 (f).
12.   L. S. B. Leakey and M. D. Leakey 1964 (OH16).   M. D. Leakey 1971 (OH30).
13.   P. V. Tobias.
14.   ————
15.   ————
16.   P. V. Tobias and G. H. R. von Koenigswald 1964, A comparison between the Olduvai
      hominines and those of Java and some implications for hominid phylogeny, *Nature,*
      Lond., **204** : 515–518 (OH30).   L. S. B. Leakey 1966, *Homo habilis, Homo erectus*
      and the australopithecines, *Nature, Lond.,* **209** : 1279–1281 (OH16).

1.   FLK West.
2.   Erosion slope about 55 m (175 ft) SW of site FLK.   2° 59′ S, 35° 20′ E.
3.   E. Kandini, assistant to M. D. Leakey, 1969.
4.   Surface of Upper Bed I, probably *ex* same horizon as OH16 and OH30 (see above, site
      FLK Maiko Gully).
6.   Lower Pleistocene.
11.   **OH26.**   lower rt M3.
12.   M. D. Leakey 1971, *Olduvai Gorge,* Vol. 3, Cambridge.
13.   P. V. Tobias, in preparation.
14.   ————
15.   ————

1.   HWK.
2.   Erosion slope at head of gully on S side of Main Gorge just E of junction with Side
      Gorge.   2° 59′ S, 35° 20′ E.
3.   An assistant to L. S. B. Leakey, 1959.
4.   Surface of Lower Bed II.
6.   Lower Pleistocene.
11.   **OH20.**   proximal diaphysis and neck of lt femur.
12.   M. H. Day 1969, Femoral Fragment of a robust australopithecine from Olduvai Gorge,
      Tanzania, *Nature, Lond.,* **221** : 230–233.
13.   M. H. Day 1969.   *Australopithecus boisei.*
14.   ————
15.   M. H. Day 1969.

1. HWK EE.
2. Living floor about 365 m (400 yds) E of site HWK.   2° 59′ S, 35° 20′ E.
3. An assistant to M. D. Leakey, 1972.
4. *In situ* in base of Sandy Conglomerate in Lower Middle Bed II.
6. Lower Pleistocene.
7. Developed Oldowan A.   M. D. Leakey 1971, *Olduvai Gorge*, Vol. 3, Cambridge.
8. *Libytherium olduvaiensis, Gazella, Antidorcas, Parmularius angusticornis, Connochaetes* cf. *africanus*.
11. **OH41.**   upper lt M1 or M2.
12. —————
13. P. V. Tobias, in preparation.
14. —————
15. —————
16. —————

1. LLK.
2. Erosion slope on N side of Side Gorge about 1 km W of junction with Main Gorge. 2° 59′ S, 35° 20′ E.
3. L. S. B. Leakey, 2 December 1960.
4. Recovered from the surface of Bed I, presumed fallen from deposit with matching matrix in Upper Bed II.
6. Lower Pleistocene.
11. **OH 9.**   calvaria (f).   **Holotype** of *Homo leakeyi* Heberer, 1963.   Plate 19.
12. L. S. B. Leakey 1961, New Finds at Olduvai Gorge, *Nature, Lond.,* **189** : 649–650.
13. —————
14. —————
15. L. S. B. Leakey 1961.
16. J. R. Napier and J. S. Weiner 1961, Olduvai Gorge and Human origins, *Antiquity,* **36** : 41–47. G. Heberer 1963, Uber einen neuen archanthropien Typus aus der Oldoway-Schlucht, *Z. Morph. Anthrop.,* **53** : 171–177. G. Kurth 1965, Die (Eu) Homininen, *in* G. Heberer (ed.) 1965, *Menschliche Abstammungslehre,* Stuttgart : 357. P. V. Tobias 1965a, Early Man in East Africa, *Science N.Y.,* **149** : 22–33. P. V. Tobias 1965b, New discoveries in Tanganyika: their bearing on hominid evolution, *Curr. Anthrop.,* **6** : 391–411. R. L. Holloway Jr 1973, New endocranial values for Olduvai and East Rudolf hominids, *Nature, Lond.,* **243** : 97–99.

1. MNK.
2. Erosion slope and on and above living floor on S side of Side Gorge about 1·5 km W of junction with Main Gorge.   2° 59′ S, 35° 19′ E.
3. N. Mbuika, assistant to L. S. B. Leakey, October 1963 (OH13); assistants to L. S. B. Leakey, 1963 (OH14, 15); P. Nzube, assistant to M. D. Leakey, 1969 (OH32).

4. *In situ* in clayey tuff 7·5 m above top of Tuff IF at top of Bed I (OH13); in re-worked tuff 8·5 m above Tuff IF (OH15); surface, inferred same level as OH13 (OH14); surface, Middle Bed II (OH32).   M. D. Leakey 1971.
6. Lower Pleistocene.
7. Oldowan.   M. D. Leakey 1971, *Olduvai Gorge,* Vol. 3, Cambridge (OH13-15).   D. N. Stiles, R. L. Hay and J. R. O'Neil 1974, The MNK Chert Factory Site, Olduvai Gorge, Tanzania, *Wld Archaeol.,* **5** : 285–308.
8. *Elephas recki* stage 3, *Ceratotherium simum, Mesochoerus olduvaiensis.*   M. Leakey 1971, *in* M. D. Leakey 1971 (OH13–15).
9. ───────
10. ───────
11. **OH13.**   juvenile: calotte (ff); palatinum (f), rt P3–M3, lt P3, M1–M3; corpus mandibulae with complete dentition.
     OH14.   juvenile: rt parietale (ff); frontale (ff).
     OH15.   adult ? male: upper rt C, M3, lt M3.
     OH32.   upper rt dm1 (?).
12. L. S. B. Leakey and M. D. Leakey 1964, Recent discoveries of fossil hominids in Tanganyika: at Olduvai and near Lake Natron, *Nature, Lond.,* **202** : 3–5 (OH13–15).   M. D. Leakey 1971 (OH32).
13. L. S. B. Leakey, P. V. Tobias and J. R. Napier 1964, A new species of the genus *Homo* from Olduvai Gorge, *Nature, Lond.,* **202** : 5–7 (OH13).   *Homo habilis.*   P. V. Tobias, in preparation.
14. ───────
15. ───────
16. P. V. Tobias and G. H. R. von Koenigswald 1964, A comparison between the Olduvai hominines and those of Java and some implications for .hominid phylogeny, *Nature, Lond.,* **204** : 515–518 (OH13).   J. T. Robinson 1965, *Homo 'habilis'* and the australopithecines, *Nature, Lond.,* **205** : 121–124 (OH13).   P. V. Tobias 1965a, Early man in East Africa, *Science, N.Y.,* **149** : 22–33.   P. V. Tobias 1965b, New discoveries in Tanganyika, their bearing on hominid evolution, *Curr. Anthrop.,* **6** : 391–411 (OH13).   R. L. Holloway Jr 1973, New endocranial values for Olduvai and East Rudolf hominids, *Nature, Lond.,* **243** : 97–99.

1. SC.
2. Erosion slope on N side of Side Gorge about 2 km W of junction with Main Gorge. 2° 59′ S, 35° 19′ E.
3. M. Mutala, assistant to M. D. Leakey, 1970 (OH36); assistant to M. D. Leakey, 1971 (OH38).
4. Surface, Upper Bed II (OH36); *in situ* in Tuff IID, Upper Bed II (OH38).
6. Lower Pleistocene.
7. Indeterminate.
8. *Elephas recki* stage 3, *Pelorovis oldowayensis, Libytherium olduvaiensis, Ceratotherium simum* (OH38).
11. **OH36.**   rt ulna.
     OH38.   lower rt M2, lt and rt upper I1.

12. ————
13. M. H. Day, in preparation (OH36).   P. V. Tobias, in preparation (OH38).
14. ————
15. ————

BED III/IV

1. JK.
2. Excavations on N side of Main Gorge about 2 km E of junction with Side Gorge.   2° 59′ S, 35° 21′ E.
3. An assistant to M. D. Leakey, 1969 (OH29); an assistant to M. Kleindienst, 1962 (OH34).
4. *In situ* in top of Bed III (OH29); *in situ* in sand below OH29 (OH34).   M. R. Kleindienst 1973, Excavations at site JK2, Olduvai Gorge, Tanzania, 1961–1962: The Geological Setting, *Quaternaria,* **17** : 145–208.
6. Lower Pleistocene.   M. R. Kleindienst 1973.
7. Acheulian (OH34).   M. R. Kleindienst 1973.
8. *Elephas recki,* stage 3, *Afrochoerus nicoli* (OH34).   M. R. Kleindienst 1973.
9. OH34 femur: $F = 1.83\%$, $P205/100F = 8.22$, $U = 6$ ppm ($c.$ 10 ppm eU308).   Animal bones: $F = 1.53–2.69\%$, $100F/P205 = 8.41–10.66$, $eU308 = 8–73$ ppm.   M. H. Day and T. I. Molleson 1976, The Puzzle from JK2– a Femur and a Tibial Fragment from Olduvai Gorge, Tanzania, *Jl. hum. Evol.,* **5** : 319–329.
11. **OH29.** M.
    OH34.   femur diaphysis (f), tibia diaphysis (f).
12. M. D. Leakey 1971, *Olduvai Gorge,* Vol. 3, Cambridge.
13. M. H. Day and T. I. Molleson 1976.
14. ————
15. M. H. Day and T. I. Molleson 1976.

1. MNK.
2. Erosion slope.   2° 59′ S, 35° 19′ E.
3. M. D. Leakey, 1935.
4. Surface, base of Bed IV.
6. Lower Pleistocene.
7. Inferred Acheulian.
9. Parietale EM550: $F = 1.95\%$, $100F/P205 = 11.9$, $eU308 = 10$ ppm, $N = $ nil.
    Parietale EM551: $F = 3.1\%$, $100F/P205 = 10.2$, $eU308 = 37$ ppm, $N = $ nil.
10. ————
11. **OH 2.**   2 fragments parietale.
12. L. S. B. Leakey 1951, *Olduvai Gorge,* Cambridge.   Cf. *Homo erectus.*
13. ————

14. ————
15. ————
16. P. V. Tobias 1965, New discoveries in Tanganyika: their bearing on hominid evolution, *Curr. Anthrop.,* **6** : 391–411.
17. British Museum (Natural History), Cromwell Road, London SW7 5BD, England.   Reg. Nos. EM550, 551.
18. British Museum (Natural History), Cromwell Road, London SW7 5BD, England.

1. VEK.
2. Erosion slope on S side of Side Gorge just W of junction with Main Gorge.   2° 59′ S, 35° 20′ E.
3. M. Leakey, 1962.
4. Surface, *ex* Bed IV.
6. Lower Pleistocene.
11. **OH12.**   calotte (f), lt palatinum with P3–M2 (all f).
12. L. S. B. Leakey and M. D. Leakey 1964, Recent discoveries of fossil hominids in Tanganyika: at Olduvai and near Lake Natron, *Nature, Lond.,* **202** : 3–5. *Homo erectus.*
13. ————
14. ————
15. ————
16. P. V. Tobias 1965, New discoveries in Tanganyika: their bearing on hominid evolution, *Curr. Anthrop.,* **6** : 391–411.   R. L. Holloway Jr 1973, New endocranial values for Olduvai and East Rudolf hominids, *Nature, Lond.,* **243** : 97–99.

1. WK.
2. Living floor on S side of Main Gorge about 1·75 km E of junction with Side Gorge.   2° 59′ S, 35° 21′ E.
3. Assistant to M. D. Leakey, 1970.
4. *In situ* in Upper Bed IV.   M. D. Leakey 1971, Discovery of Postcranial remains of *Homo erectus* and associated artefacts in Bed IV at Olduvai Gorge, Tanzania, *Nature, Lond.,* **232** : 380–383.
6. Middle Pleistocene.
7. Acheulian.   M. D. Leakey 1971.
11. **OH28.**   lt femur diaphysis (f), lt os coxae (f).
12. M. H. Day 1971, Postcranial remains of *Homo erectus* from Bed IV, Olduvai Gorge, Tanzania, *Nature, Lond.,* **232** : 383–387.
13. M. H. Day 1971.   *Homo erectus.*
14. ————
15. M. H. Day 1971.

1.   Site unnamed.
2.   Erosion slope between sites VEK and MNK.   2° 59′ S, 35° 20′ E.
3.   M. Mwoka, assistant to M. D. Leakey, 1968.
4.   Surface of indeterminate gravel, probably *ex* Bed IV, although just possibly Bed III, on basis of matrix adhering to specimen.
6.   Middle or Lower Pleistocene.
11. **OH22.**   rt corpus mandibulae, P3–M2, roots of I1–C and M3.
12. M. D. Leakey 1969, Recent discoveries of hominid remains at Olduvai Gorge, Tanzania, *Nature, Lond.*, **223** : 756.   Cf. *Homo erectus.*
13. ————
14. ————
15. ————

MASEK BEDS

1.   FLK.
2.   Erosion slope on cliff above site of OH 5.   2° 59′ S, 35° 20′ E.
3.   E. Kandini, assistant to M. D. Leakey, 1968.
4.   *In situ* in Masek beds, in channel between Lower Aeolian Tuff and the Norkilili Member.
6.   Middle Pleistocene.
7.   Acheulian.
8.   *Clarias, Hippopotamus gorgops.*
11. **OH23.**   lt corpus mandibulae (f), P4–M2 (all abraded), roots P3.
12. M. D. Leakey 1969, Recent discoveries of hominid remains at Olduvai Gorge, Tanzania, *Nature, Lond.*, **223** : 756.
13. ————
14. ————
15. ————

NDUTU BEDS (LOWER)

1.   DK.
2.   Stream gully just W of DK.   2° 59′ S, 35° 22′ E.
3.   M. Matumbo, assistant to L. S. B. Leakey, 1962.
4.   Surface.   Source unknown but matrix suggests Lower Ndutu Beds.   M. D. Leakey, R. L. Hay, D. L. Thurber, R. Protsch and R. Berger 1972, Stratigraphy, archaeology and age of the Ndutu and Naisiusiu Beds, Olduvai Gorge, Tanzania, *Wld Archaeol.*, **3** : 328–341.
6.   Middle Pleistocene.
11. **OH11.**   lt palatinum, roots C–M2.
12. P. V. Tobias 1965, New discoveries in Tanganyika: their bearing on hominid evolution, *Curr. Anthrop.*, **6** : 391–411.

13. ———————
14. ———————
15. ———————

NAISIUSIU BEDS

1. RK.
2. Burial on N side of Main Gorge about 4·6 km E of junction with Side Gorge.   2° 59′ S, 35° 22′ E.
3. H. Reck, 1913.
4. Top of Bed II.   H. Reck 1914a, Erste vorläufige Mitteilung über den Fund eines fossilen Menschenskelets aus Zentralafrika, *Sber. Ges. naturf. Freunde Berl.*, **3** : 81–95.
5. Ultra-contracted burial.   H. Reck 1914a.   Burial into Bed II.   P. G. H. Boswell 1932, The Oldoway human skeleton, *Nature, Lond.*, **130** : 237–238.   Burial, inferred from same horizon as Naisiusiu Beds because dates from eggshell and collagen in mammal bones from Naisiusiu Beds at Second Fault agree well with date for OH 1 skeleton (see section 10).
6. Naisiusiu Beds are Late Pleistocene.   M. D. Leakey, R. L. Hay, D. L. Thurber, R. Protsch and R. Berger 1972, Stratigraphy, archaeology and age of the Ndutu and Naisiusiu Beds, Olduvai Gorge, Tanzania, *Wld Archaeol.*, **3** : 328–341.
7. ———————
8. Extant fauna.   L. S. B. Leakey 1951, *Olduvai Gorge,* Cambridge University Press.
9. OH 1 bone: $F = 2·7\%$, $100F/P205 = 10·2$, $eU308 = 2$ ppm.   Mammalian bones from Bed II: $eU308 = 30$–1280 ppm.
10. A1: $16{,}920 \pm 920$ BP (UCLA-1740) on basis of C14 dating of collagen of bone from OH 1.
     A3: 17,000 BP on basis of C14 dating of ostrich eggshell from type section of Naisiusiu Beds, and of collagen content of mammal bones from same section $17{,}550 \pm 1{,}000$ BP (UCLA-1695).   M. D. Leakey *et al.* 1972.
11. **OH 1.**   adult male: skeleton (i).
12. H. Reck 1914a.
13. W. Gieseler and T. Mollison 1929, Untersuchungen über den Oldoway Fund, *Verh. Ges. phys. Anthrop.*, **3** : 50–67.
14. L. S. B. Leakey 1951.
15. H. Reck 1914a.   W. Gieseler and T. Mollison 1929.   R. Protsch 1974, The Age and Stratigraphic Position of Olduvai Hominid I, *J. hum. Evol.*, **3** : 379–385.
16. H. Reck 1914b, Zweite vorläufige Mitteilung über fossile Tier- und Menschenfunde aus Oldoway in Zentralafrika, *Sber. Ges. naturf. Freunde Berl.*, **7** : 305–318.   C. Foster Cooper and D. M. S. Watson 1932, The Oldoway human skeleton, *Nature, Lond.*, **192** : 312–313.   L. S. B. Leakey 1928, The Oldoway skull, *Nature, Lond.*, **121** : 499–500.   L. S. B. Leakey 1935, *The Stone Age Races of Kenya,* Oxford University Press.   L. S. B. Leakey, H. Reck, P. G. H. Boswell, A. T. Hopwood and J. D. Solomon 1933, The Oldoway human skeleton, *Nature, Lond.*, **131** : 397–398.
17. Institut für Anthropologie und Humangenetik, Ludwig Maximilians Universität, Richard Wagner Strasse, 10/1, München, Federal Republic of Germany.

18. Institut für Anthropologie und Humangenetik, Ludwig Maximilians Universität, Richard Wagner Strasse, 10/1, München, Federal Republic of Germany.

UNKNOWN HORIZON

1. FLK, Maiko Gully.
2. In stream bed about 180 m (200 yds) S of site FLK.   2° 59′ S, 35° 20′ E.
3. An assistant to L. S. B. Leakey, 1963.
4. Surface.
11. **OH17.**   dm.
12. M. D. Leakey 1971, *Olduvai Gorge,* Vol. 3, Cambridge.
13. ————
14. ————
15. ————

1. FLK North.
2. On erosion slope about 180 m (200 yds) N of site FLK.   2° 59′ S, 35° 20′ E.
3. P. Nzube, assistant to M. D. Leakey, 1968.
4. Surface.
11. **OH21.**   upper lt M1.
12. M. D. Leakey 1969, Recent discoveries of hominid remains at Olduvai Gorge, Tanzania, *Nature, Lond.,* **223** : 756.
13. ————
14. ————
15. ————

1. Geologic locality 54.
2. Erosion slope on S side of Main Gorge about 8·5 km W of junction with Side Gorge.   2° 57′ S, 35° 16′ E.
3. P. Nzube, assistant to M. D. Leakey, 1968.
4. Surface.
11. **OH25.**   lt parietale (f in region of asterion).
12. M. D. Leakey 1971, *Olduvai Gorge,* Vol. 3, Cambridge.
13. ————
14. ————
15. ————

PENINJ

1. Peninj (Natron).
2. Erosion gully site (RHS) by Peninj river, W of Lake Natron.   2° 12′ S, 35° 56′ E.
3. K. Kimeu, 11 January 1964.   Leaders G. Ll. Isaac and R. E. Leakey.
4. Lake delta sandstones of Humbu Formation, Peninj Group.   L. S. B. and M. D. Leakey 1964, Recent Discoveries of Fossil Hominids in Tanganyika: at Olduvai and near Lake Natron, *Nature, Lond.,* **202** : 3–5.   Lower part of Peninj Group.   G. Ll. Isaac 1967, The Stratigraphy of the Peninj Group, *in* W. W. Bishop and J. D. Clark (eds) 1967, *Background to Human Evolution in Africa,* Chicago : 229–257.
5. ————
6. Lower Pleistocene, equivalent to upper part of Olduvai Bed II.   L. S. B. and M. D. Leakey 1964.   L. S. B. Leakey, P. V. Tobias and J. R. Napier 1964, A New Species of the Genus *Homo* from Olduvai Gorge, *Nature, Lond.,* **202** : 5–7.
7. Apart from doubtful quartz fragment no artifacts associated.   Lower Acheulian industry, mainly in lava but including a few quartz cores, occurs in upper part of Humbu Formation.   G. Ll. Isaac 1967 : 245–252.
8. Lower Pleistocene fauna including *Metridiiochoerus andrewsi, Damaliscus antiquus, D. niro, Elephas recki, Alcelaphus howardi, Phenacotragus* cf. *recki, Libytherium.*   G. Ll. Isaac 1967 : 245.
9. ————
10. A2: 1·4–1·6 Myr on K/Ar dating of basalt flow just above level of jaw.   G. H. Curtis 1967, Notes on some Miocene to Pleistocene Potassium/Argon Results, *in* W. W. Bishop and J. D. Clark (eds) 1967 : 367.
11. **Peninj 1.**   young adult: mandibula, complete dentition.
12. L. S. B. and M. D. Leakey 1964.   *Australopithecus (Zinjanthropus).*
13. P. V. Tobias 1965, The early *Australopithecus* and *Homo* from Tanzania, *Anthropologie, Prague,* **3** : 43–48.
14. ————
15. L. S. B. and M. D. Leakey 1964.
16. ————
17. To be housed in the National Museum, P.O. Box 511, Dar es Salaam.
18. ————

TUNISIA

There are no Pleistocene hominid remains known from Tunisia.   The earliest known are Capsian burials, and these have been listed both in *CHF* 1953, and in L. Balout 1955, *Les Hommes Préhistoriques du Maghreb et du Sahara,* Alger.
The following sites should be noted:

      Bir Hamairia
      Clariond
      El-Mekta

ZAIRE

Jean de HEINZELIN de Braucourt

Geologisch Instituut,
Universiteit Gent,
Belgium.

&

François TWIESSELMANN

Institut Royal des Sciences Naturelles de Belgique,
Bruxelles,
Belgium.

ISHANGO

1. Ishango.
2. Terraces on right bank of Semliki River, just before its outlet from Lake Edward. 0° 08′ S, 29° 36′ E.
3. H. Damas, 4 December 1935 (Ishango 1); J. de Heinzelin, 25 April–23 July 1950 (Ishango 2–8).
4. Tuffaceous layers of river terrace; attributed to Principal Fossiliferous Layer (Ishango 1). Terrace Tt (10·5 m above lake level), Principal Fossiliferous Layer (N.F.PR.) = kitchen-midden debris in lake shore deposits overlain by lacustrine tuffite; Terrace Tp (12 m above lake level), Post-Emersion Zone (Z.P–E.), kitchen-midden debris on surface of terrace gravels, overlain by colluvium (Ishango 2–8). J. de Heinzelin 1955, Le Fossé Tectonique sous le Parallèle d'Ishango, *Explor. Parc. natn. Albert Miss. J. de Heinzelin de Braucourt,* **1** : 63–65.
5. ───────
6. Epi-Pleistocene, Makalian. J. de Heinzelin 1955 : 9.
7. N.F.PR.: Early Mesolithic, ' Civilisation d'Ishango ' with stone and bone artifacts including harpoons with double row of barbs, grindstones, no pottery. J. de Heinzelin 1957, Les Fouilles d'Ishango, *Explor. Parc natn. Albert Miss. J. de Heinzelin de Braucourt,* **2** : 20–77 and plates. P-E.Z.: Mesolithic (cf. Smithfield) with stone and bone artifacts but no harpoons reported, no pottery. J. de Heinzelin 1955 : 134–135; J. de Heinzelin 1957 : 20–77.
8. N.F.PR.: *Lutra maculicollis, Thryonomys swinderianus, Potamochoerus porcus, Hippopotamus amphibius.* Aves: *Struthio camelus massaicus;* Pisces; Mollusca. Z.P-E.: *Colobus, Panthera pardus, Hippopotamus amphibius;* Aves: *Plateles alba;* Pisces. P. H. Greenwood, V. Verheyen, A. T. Hopwood and X. Misonne 1959, *Explor. Parc natn. Albert Miss J. de Heinzelin de Braucourt,* **4.** J. de Heinzelin 1955 : 64–65.
9. Human femur (N.F.PR.): F $= 2·1\%$, 100F/P205 $= 7·1$, eU308 $= 16$ ppm, N $=$ nil. *Hippopotamus* bone (N.F.PR.): F $= 1·7\%$, 100F/P205 $= 6·6$, eU308 $= 64$ ppm, N $=$ nil.
10. N.F.PR. $= c.$ 9000 BP. J. de Heinzelin 1962, Ishango, *Scient. Am.,* **206** : 106. Correlation with dated volcanic ash layers in deposits of Mahoma Lake (Ruwenzori) and comparable industry in Turkana, suggests 4700–7500 BP. J. de Heinzelin 1971, *in lit.*
11. **Ishango 1.**   adult: rt ramus mandibulae (ff), M2.

    Ishango 2, 3, 4.   (N.F.PR.) adult: 10 cranial fragments, including frontale (ff), lt parietale (ff), rt parietale (ff), mandibula (f), dentes, isolated upper lt M1. Also: rt ramus mandibulae (ff), dentes; rt ramus mandibulae (ff), dens; atlas, vertebra cervicalis lt humerus (f); rt juvenile humerus (f); 2 lt, 1 rt ulnae; 3 rt radii; 1 (i), 2 femora (f); lt patella; rt tibia; lt talus; juvenile rt cuboideum, 2 metacarpalia, 20 metatarsalia, 9 phalanges manus, 3 phalanges pedis.

    Ishango 5, 6, 7, 8.   and unnumbered individuals. (Z.P-E). 3 fragments of parietale; mandibula without condyles (f), dentes; rt ramus mandibulae (f), dentes; 4 rami mandibulae (ff), dentes; lt ramus mandibulae (ff), dentes; rt ramus mandibulae with symphysis (ff), dentes; lt ulna; lt radius (ff); 3 patellae; upper epiphyses lt and rt tibiae; lt talus; lt calcaneus, 4 metacarpalia, 2 phalanges manus, 1 phalanx pedis.

12.   J. de Heinzelin 1955.
13.   F. Twiesselmann 1958, Les ossements humains du gîte mésolithique d'Ishango, *Explor. Parc natn. Albert Miss. J. de Heinzelin de Braucourt,* **5** : 3–123.
14.   ————
15.   F. Twiesselmann 1958.
16.   H. Damas 1940, Observations sur des couches fossilifères bordant la Semliki, *Revue Zool. Bot. afr.,* **33** : 265–272.
17.   Institut Royal des Sciences Naturelles de Belgique, rue Vautier, 31, Bruxelles, Belgium.
18.   Institut Royal des Sciences Naturelles de Belgique, rue Vautier, 31, Bruxelles, Belgium.

KAKONTWE   10° 59′ S, 26° 42′ E.

In the NE corner of a quarry at Kakontwe, Katanga, A. Anciaux de Faveaux and M. Hedo excavated 2 lt molar teeth of a child aged 4–5 years from a bone breccia bed (III).   The associated fauna is recent, the industry Stillbay.   The age is Final Upper Pleistocene.

A. Anciaux de Faveaux 1957, Les Brèches ossifères de Kakontwe, *Proc. 3rd Pan-Afr. Congr. Prehist*: 98–101.   F. Twiesselman 1965, Déscription de deux dents molaires humaines d'âge Paléolithique supérieur, provenant de la brèche ossifère de Kakontwe (Katanga), *Bull. Soc. belge Anthrop. Prehist.,* **75** : 107–119.

The teeth are preserved at the Musée Leopold II, Lubumbashi, Katanga.

ZAMBIA

David Walter PHILLIPSON,

British Institute in Eastern Africa,
P.O. Box 30710,
Nairobi,
Kenya.

&

John Desmond CLARK,

Department of Anthropology,
University of California,
Berkeley, California 94720,
U.S.A.

BROKEN HILL

1. Broken Hill.
2. Cave in No. 1 Kopje (later destroyed by quarrying), Broken Hill Mine, Kabwe, Central Province.   14° 27′ S, 28° 26′ E.
3. T. Zwigelaar, 17 June 1921 (cranium); A. S. Armstrong and Mrs A. W. Whittington, 1921; A. Hrdlička, 1925 (other remains).
4. Cranium filled with secondary zinc ore, but within lead-impregnated zone of breccia.   J. D. Clark, K. P. Oakley, L. H. Wells and J. A. C. McClelland 1950, New Studies on Rhodesian Man, *Jl. R. anthrop Inst.*, **77** : 7–32.
5. ————
6. Upper Pleistocene.   J. D. Clark *et al.* 1950.
7. Early Middle Stone Age, ' Proto-Stillbay ' tradition.   J. D. Clark *in* J. D. Clark *et al.* 1950 : 13–30.   J. D. Clark 1959, Further Excavations at Broken Hill, Northern Rhodesia, *Jl. R. anthrop. Inst.*, **89** : 201–232.   ' Proto-Stillbay ' = facies of Sangoan or Late Acheulian.
8. Mammalian fauna mainly of extant species including the rare *Litocranius* cf. *walleri,* but seven extinct genera are present, including *Homoioceras baini.*   L. S. B. Leakey, *in* J. D. Clark 1959 : 225–230.   (' Bat bones ' proved to be mainly shrew, *Crocidura).*   R. G. Klein 1973, Geological Antiquity of Rhodesian Man, *Nature, Lond.,* **244** : 311–312.
9. Broken Hill 1 cranium: $F = 1\cdot8\%$, $100F/P205 = 6\cdot1$, $eU308 = 11$ ppm, $N = 0\cdot1\%$.   Broken Hill 2 maxilla: $F = 0\cdot7\%$, $100F/P205 = 6\cdot1$, $eU308 = 20$ ppm, $N = 0\cdot1\%$.   Broken Hill 3 parietal: $F = 1\cdot2\%$, $100F/P205 = 4\cdot3$, $eU308 = 48$ ppm, $N = $ nil.   Rt femur (E907): $F = 1\cdot0\%$, $100F/P205 = 4\cdot3$, $eU308 = 32$ ppm, $N = 0\cdot1\%$.   Lt femur (EM793): $F = 0\cdot8\%$, $100F/P205 = 2\cdot8$, $eU308 = 32$ ppm, $N = $ nil.   Faunal controls.   *Crocidura:* $F = 1\cdot5\%$, $100F/P205 = 4\cdot9$, $eU308 = 30$ ppm, $N = $ nil.   15 other bones: $N = 0–0\cdot26\%$.   K. P. Oakley 1957, The Dating of Broken Hill (Rhodesian) Man, *in* G. H. R. von Koenigswald (ed.) 1957, *Hundert Jahre Neanderthaler,* Utrecht : 265–269.
10. A1: 110,000 BP on basis of racemization of aspartic acid in femur shaft fragment EM793.   J. L. Bada, R. A. Schroeder, R. Protsch and R. Berger 1974, Concordance of Collagen-Based Radiocarbon and Aspartic-Acid Racemization Ages, *Proc. Nat. Acad. Sci. U.S.A.,* **71** : 914–917.

    A3: perhaps *c.* 45,000–35,000 BP on basis of C14 dating of charcoal associated with industries of the Sangoan to Proto-Stillbay traditions in Pomongwe Cave, Matopo Hills, Rhodesia.   ' Stillbay ': 35,530 $\pm$ 780 BP (SR-39); ' Proto-Stillbay ' 42,200 $\pm$ 2300 BP (SR-8); ' Proto-Stillbay ' > 42,000 BP (SR-7).   37,900–46,100 BP on basis of C14 dating of carbon associated with Sangoan Industrial complex at Kalambo Falls, Zambia.
11. **Broken Hill 1.**   adult male: cranium (E686); all maxillary dentes present and worn, except rt I2; rt M3 roots only.   **Holotype** of *Homo rhodesiensis* Woodward, 1921.   **Holotype** of *Cyphanthropus rhodesiensis* Pycraft, 1928.   Plate 20.

    Broken Hill 2.   rt maxilla (f) (E687).

    Broken Hill 3.   parietale (f) (E897).

    Broken Hill post-cranial bones: at least 3 individuals not in clear association with numbered individuals: sacrum (i) (E688), lt ilium, female (f) (E726), rt ilium and ischium, male (ff) (E719), rt humerus (part of distal end) (E898), lt femur (part of proximal end) (E689), lt femur (part of distal end) (E689), lt femur, female? (part of diaphysis) (E690), lt(?) femur (diaphysis fragment) (EM793), rt femur (part of proximal end) (E907), lt tibia (i) (E691).

12. A. S. Woodward 1921, A New Cave Man from Rhodesia, South Africa, *Nature, Lond.,* **108** : 371–372. *Homo rhodesiensis.* J. D. Clark, D. R. Brothwell, R. Powers and K. P. Oakley 1968, Rhodesian Man: Notes on a New Femur Fragment, *Man,* **3** : 105–111 (EM793).

13. W. P. Pycraft *et al.* 1928, Rhodesian Man and Associated Remains, *British Museum (Natural History),* London, 75pp.

14. W. E. Le Gros Clark 1928, Rhodesian Man, *Man,* **28** : 206–207. L. H. Wells 1950, *in* J. D. Clark *et al.* 1950 : 11–12.

15. W. P. Pycraft *et al.* 1928. J. D. Clark *et al.* 1968 (EM793).

16. G. M. Morant 1928, Studies of Palaeolithic Man, III. The Rhodesian Skull and its Relationship to Neanderthaloid and Modern Types, *Ann. Eugen.,* **3** : 337–360. R. Singer 1958, The Rhodesian, Florisbad and Saldanha Skulls, *in* G. H. R. von Koenigswald (ed.) 1958, *Hundert Jahre Neanderthaler,* Utrecht: 52–62, pl. 21. A. Hrdlička 1930, The Skeletal Remains of Early Man, *Smithson. misc. Collns.,* **83** : 98–144. R. G. Klein 1970, Problems in the study of the Middle Stone Age of South Africa, *S. Afr. archaeol. Bull.,* **25** : 127–135. G. P. Rightmire 1976, Relationships of Middle and Upper Pleistocene hominids from sub-Saharan Africa, *Nature, Lond.,* **260** : 238–240.

17. British Museum (Natural History), Cromwell Road, London SW7 5BD, England.

18. British Museum (Natural History), Cromwell Road, London SW7 5BD, England. Wenner-Gren Foundation for Anthropological Research, 14 East 71st Street, New York, N.Y., 10021, U.S.A.

CHIPONGWE   15° 40′ S, 28° 14′ E.

In 1930, R. A. Dart and an Italian Scientific Expedition excavated a cave at Chipongwe, 0·5 km W of the Great North Road, 34 km S of Lusaka, Central Province. A cranium (f) and fragments of other individuals were found in the scree of the cave deposits. The remains were associated with a Late Stone Age industry (cf. Wilton) and are probably dated *c.* 4000 BP (A3).

J. D. Clark 1950, *Stone Age Cultures of Northern Rhodesia,* Cape Town: 143–150. J. D. Clark and M. J. Toerien 1955, Human Skeletal and Cultural Material from a deep cave at Chipongwe, Northern Rhodesia, *S. Afr. archaeol. Bull.,* **10** : 107–116.

GWISHO

1. Gwisho A.

o

2.   Site A, Gwisho Hotsprings, Lochinvar National Park, Monze District, Southern Province.   15° 59′ S, 27° 14′ E.
3.   Excavations by C. Gabel, 1960–1961.
4.   Calcareous spring deposits.   C. Gabel 1965, *Stone Age Hunters of the Kafue,* Boston.
5.   Shallow burials within the spring deposits.
6.   Holocene.   C. Gabel 1965.
7.   Late Stone Age, cf. Wilton.   C. Gabel 1965.
8.   Extant fauna.   C. Gabel 1965.
9.   Hominid mandibula: F = 2·1%, 100F/P205 = 5·7, eU308 = 29 ppm, N = 0·1%.
10.   A2: 5000–4000 BP.   Duration of occupation of Gwisho A indicated by four C14 dates: depth 1·5 m 4700 ± 100 BP (UCLA-174); depth 2·4 m 4230 ± 100 BP and 4350 ± 150 BP (UCLA-173); depth 2·4 m 4450 ± 150 BP (M-1323); depth 2·4 m 4650 ± 150 BP (M-1324).
11.   **GSA 1.**   adult male (?): skeleton (ff).
       GSA 2.   adult male (?): skeleton (f).
       GSA 3.   young adult female (?): skeleton (ff).
       GSA 4.   young adult female (?): skeleton (ff).
       GSA 5.   adult: cranium (ff), mandibula (f).
       GSA 6.   adult male (?): skeleton (ff).
       GSA 7.   post-cranial skeleton (ff).
       GSA 8.   post-cranial skeleton (ff).
       GSA 9.   adult male (?): skeleton (ff).
       GSA 10.   adult male (?): skeleton (ff).
       GSA 11.   young adult female (?): skeleton (ff).
       GSA 12.   adult female (?): skeleton (ff).
       GSA 13.   young adult male (?): skeleton (ff).
       GSA 14.   adult male (?): skeleton (ff).
       GSA 15.   mandibula (f).
       GSA 16.   cranial fragments, 6 dentes.   Possibly same individual as GSA 15.
12.   C. Gabel 1962, Human Crania from the Later Stone Age of the Kafue Basin, Northern Rhodesia, *S. Afr. J. Sci.,* **58** : 307–314.
13.   C. Gabel 1965.
14.   D. R. Brothwell 1971, *in* B. M. Fagan and F. L. van Noten 1971, *The Hunter-Gatherers of Gwisho,* Tervuren : 37–47.
15.   C. Gabel 1965.
16.   C. Gabel 1963a, Lochinvar Mound: a Later Stone Age Campsite in the Kafue Basin, *S. Afr. archaeol. Bull.,* **18** : 40–48.   C. Gabel 1963b, Further Human Remains from the Central African Later Stone Age, *Man,* **63** : 39–43.
17.   Livingstone Museum, P.O. Box 498, Livingstone.   Reg. No. 8958.
18.   ————

1.   Gwisho B.
2.   Site B, Gwisho Hotsprings, Lochinvar National Park, Monze District, Southern Province.   15° 59′ S, 27° 14′ E.
3.   Excavations by B. M. Fagan and F. L. van Noten, 1963–1964.

4. Calcareous spring deposits.   B. M. Fagan and F. L. van Noten 1971, *The Hunter-Gatherers of Gwisho,* Tervuren.
5. Shallow burials within the spring deposits.
6. Holocene.   B. M. Fagan and F. L. van Noten 1971.
7. Later Stone Age, cf. Wilton.   B. M. Fagan and F. L. van Noten 1971.
8. Extant fauna.   B. M. Fagan and F. L. van Noten 1971.
9. —————
10. A2: 5000–3000 BP. Duration of occupation of Gwisho B indicated by three C14 dates: depth 0·45–0·47 m 3660 ± 70 BP (GrN-4305); depth 0·61–0·62 m 3680 ± 70 BP (GrN-4306); depth 1·70–1·80 m 4785 ± 70 BP (GrN-4307).
11. **GSB 1.**   juvenile: skeleton (ff).
    GSB 1A.   adult: mandibula (f).
    GSB 1A, 2. parietale (f).
    GSB 2.   juvenile: skeleton (ff).
    GSB 3.   adult male (?): skeleton (ff).
    GSB 3, 2.   adult: maxilla (f).
    GSB 4A.   juvenile: skeleton (ff).
    GSB 4B.   adult: cranium, mandibula (ff).
    GSB 5.   adult female (?): skeleton (ff).
    GSB 6.   adult female (?): skeleton (ff).
    GSB 7.   adult female (?): skeleton (ff).
    GSB 8.   juvenile female (?): skeleton (ff).
    GSB IB, 8.   infant: skeleton (ff).
    GSB 9.   adult male: skeleton (ff).
    GSB 10.   infant: skeleton (ff).
    GSB 11.   adult female: skeleton (ff).
    GSB 12.   adult male (?): skeleton (ff).
    GSB 13.   infant: skeleton (ff).
    GSB 14.   infant: skeleton (ff).
    GSB IA, 14. adult: skull (ff), I (f).
    GSB 15.   adult female (?): skeleton (ff).
    GSB 16.   juvenile: skeleton (ff).
    GSB 17.   adult male (?): skeleton (ff).
    GSB 18.   juvenile: skeleton (ff).
    GSB 19.   adult male: skeleton (f).
    GSB 20.   adult: ossa longa (ff).
12. F. L. van Noten 1964, Archeologische Opgravingen in Rhodesie 1963, *Africa-Tervuren,* **10** : 1–7.
13. D. R. Brothwell 1971, *in* B. M. Fagan and F. L. van Noten 1971 : 37–47.
14. —————
15. D. R. Brothwell 1971.
16. C. Gabel 1965, *Stone Age Hunters of the Kafue,* Boston.
17. Livingstone Museum, P.O. Box 498, Livingstone.
18. —————

An infant skeleton burial from Site C in Gwisho Hotsprings was excavated by B. M. Fagan and F. L. van Noten in 1964 and is probably of approximately similar age to Gwisho A and B remains.

   B. M. Fagan and F. L. van Noten 1971, *The Hunter-Gatherers of Gwisho:* 37–47, Tervuren.

KABWE MAN

See under **BROKEN HILL**

KALEMBA

1. Kalemba.
2. Kalemba rockshelter, SE side of Cipwete valley, 2 km NE of Mbangombe village, Chadiza District, Eastern Province.   14° 07′ S, 32° 30′ E.
3. Excavations by D. W. Phillipson, 1971.
4. Stratified cave deposits.   Horizon L (Kalemba 1), horizon O (Kalemba 2–4), horizon Q (Kalemba 5).   D. W. Phillipson 1976, *The Prehistory of Eastern Zambia,* Nairobi.
5. Shallow grave dug in horizon L from horizon O (Kalemba 2, 3).   Placed under stone slab (Kalemba 4).   Shallow grave within horizon Q (Kalemba 5).   D. W. Phillipson 1976.
6. Late Pleistocene/Holocene.   D. W. Phillipson 1976.
7. Late Stone Age, cf. Nachikufan I (Kalemba 1), Makwe Industry (Kalemba 2–5).   D. W. Phillipson 1976.
8. Extant fauna.   J. M. Harris 1976, *in* D. W. Phillipson 1976 : 161–162.
9. ————
10. Kalemba 1.   A2: *c.* 15,000 BP on basis of C14 dates of bone apatite from horizon K: 14,800 ± 1000 BP (GX-2766) and from horizon N: 15,330 ± 1100 BP (GX-2769).
    Kalemba 2–4.   A2: *c.* 8000–7000 BP on basis of C14 dates from horizon O of charcoal: 6810 ± 300 BP (N-1388) and of bone apatite: 7915 ± 300 BP (GX-2770).
    Kalemba 5.   A2: *c.* 5000–4500 BP on basis of C14 dates of charcoal from horizon Q: 5040 ± 110 BP (N-1386) and 4480 ± 90 BP (N-1387).
11. **Kalemba 1.**   I (f).
    Kalemba 2.   young adult female (?): cranium (f), mandibula, 5 vertebrae (f).
    Kalemba 3.   infant: cranium (ff), 2 ribs (ff), 4 vertebrae (ff).
    Kalemba 4.   juvenile: cranium (f).
    Kalemba 5.   adult male (?): cranium (f), mandibula, postcranial fragments.
12. D. W. Phillipson 1976.
13. H. de Villiers 1976, *in* D. W. Phillipson 1976: 163–165.
14. ————
15. H. de Villiers 1976.
16. ————
17. Livingstone Museum, P.O. Box 498, Livingstone.
18. ————

LEOPARD'S HILL

1. Leopard's Hill.
2. Cave in S side of limestone ridge 1 km W of Leopard's Hill Ranch-house which lies 52 km SW of Lusaka, Central Province.   15° 36' S, 28° 44' E.
3. R. P. Odendaal and J. D. Clark, 1946.
4. Recovered from spoil of commercial guano excavations (Leopard's Hill 1).   Recovered in situ from occupation level exposed by commercial excavation (Leopard's Hill 2).   Zambia National Monuments Commission records.
5. ————
6. Late Pleistocene/Holocene.   S. F. Miller 1972, Archaeological Sequence of the Zambian Later Stone Age, *Proc. 6th Pan-Afr. Congr. Prehist., Dakar, 1967.*
7. Late Stone Age, cf. Nachikufan.   Probably 'proto-Late Stone Age' (Leopard's Hill 2).   S. F. Miller, in preparation.
8. Faunal remains from the site are being studied by H. B. S. Cooke.   For a preliminary note see H. B. S. Cooke 1950, *in* J. D. Clark 1950, *Stone Age Cultures of Northern Rhodesia,* Cape Town : 138–141.
9. ————
10. Leopard's Hill 1.   Position uncertain in Late Stone Age sequence dated by C14 from 16,715 $\pm$ 125 BP (SR-126) to 1415 $\pm$ 125 BP (SR-126).   S. F. Miller 1971, The Age of the Nachikufan Industries in Zambia, *S. Afr. archaeol. Bull.,* **26** : 143–146.
    Leopard's Hill 2.   A2. 25,000–20,000 BP on basis of C14 dates from charcoal probably from the same layer: 23,600 $\pm$ 360 BP (UCLA-1429A), 22,600 $\pm$ 510 BP (UCLA-1429BO) 21,550 $\pm$ 950 BP (GX-957).   S. F. Miller 1972.
11. **Leopard's Hill 1.**   occipitale (f).
    Leopard's Hill 2.   parietale (ff).
12. J. D. Clark 1950.
13. L. H. Wells 1950, *in* J. D. Clark 1950 : 143–150.   Bushman type.
14. ————
15. ————
16. ————
17. Livingstone Museum, P.O. Box 498, Livingstone, Zambia.   Temporarily at University of Cape Town Medical School, South Africa.
18. ————

LOCHINVAR

See under **GWISHO**

MAKWE   14° 25' S, 31° 56' E.

In 1966, D. W. Phillipson excavated hominid remains in a rockshelter in Little Makwe Hill,

Katete District, Eastern Province, 2·5 km NNW of boundary post 16 on the Zambia/
Moçambique border.   The hominid remains come from different horizons in a stratified cave
deposit, together with an extant fauna and Late Stone Age, Makwe Industry.
Horizon 2ii yielded an incisor tooth and carries an A2 date of 4010 ± 90 BP (SR-204).
Horizon 4i-ii yielded 3M and a radius (ff) and carries an A2 date of *c.* 2000–1000 BP.
D. W. Phillipson 1976, *The Prehistory of Eastern Zambia,* Nairobi: 107–109.
The remains are preserved at the Livingstone Museum, P.O. Box 498, Livingstone.

MARAMBA   17° 53′ S, 25° 52′ E.

In 1939, J. D. Clark excavated a skeleton from an open site on the N bank of the Maramba
River near the bridge on the Livingstone – Victoria Falls road 4·5 km S of Livingstone,
Southern Province.
The eroded burial was associated with a Late Stone Age industry.   The age is estimated
(A3) at *c.* 5000 BP.
J. D. Clark 1950, *Stone Age Cultures of Northern Rhodesia,* Cape Town: 143–150.
The skeleton (f, ff) is preserved at the Medical School, University of the Witwatersrand,
Johannesburg, Hospital Street, Johannesburg, South Africa.

MUMBWA

1.   Mumbwa (1930).
2.   Cave in limestone 2·4 km W of Mumbwa, Central Province.   14° 58′ S, 27° 02′ E.
3.   Excavations by R. A. Dart, 1930 and Italian Scientific Expedition.
4.   Cave deposits.   Stratum V (Mumbwa 1), Stratum IV (Mumbwa 2–4).   R. A. Dart
     and N. del Grande 1931, The Ancient Iron-smelting Cavern at Mumbwa, *Trans. R.
     Soc. S. Afr.,* **19** : 379–427.
5.   Burials in Middle Stone Age horizons, possibly dug from higher levels.   R. A. Dart and
     N. del Grande 1931.
6.   Late Pleistocene or Holocene.   J. D. Clark 1950, *Stone Age Cultures of Northern
     Rhodesia,* Cape Town.
7.   Late Stone Age or Middle Stone Age.   R. A. Dart and N. del Grande 1931.   J. D.
     Clark 1942, Further Excavations (1939) at Mumbwa caves, Northern Rhodesia, *Trans.
     R. Soc. S. Afr.,* **29** : 133–201.
8.   Fauna of extant type but with *Equus* cf. *kuhni* in the Middle Stone Age levels.   H. B. S.
     Cooke 1950, *in* J. D. Clark 1950 : 137–142.   R. Broom 1942, *in* J. D. Clark
     1942 : 197–198.   K. H. Barnard 1942, *in* J. D. Clark 1942 : 198.
9.   ————
10. A1: 19,780 ± 130 BP (UCLA-1750C) on basis of C14 dating of collagen in bone
     of Mumbwa 1.

A2: 20,000 — 18,000 BP on basis of C14 dating of faunal remains from stratum : 20,450 ± 340 BP (UCLA-1750IV/D) and 18,000 ± 370 BP (UCLA-1750B). R. Protsch 1975, The absolute dating of Upper Pleistocene subSahara fossil hominids and their place in human evolution, *J. hum. Evol.,* **4** : 297–322.

11. **Mumbwa 1.**   calvaria (ff), clavicula (f), 2 humeri (f), ulna (f), fibula (f). More than one individual represented.

   Mumbwa 2.   cranium (ff).

   Mumbwa 3, 4*.   2 calvaria, 2 maxillae, mandibula, humerus (ff), 2 femora (ff), 2 fibulae (ff).

12. R. A. Dart and N. del Grande 1931.

13. T. R. Jones 1940, Human Skeletal Remains from the Mumbwa Cave, Northern Rhodesia, *S. Afr. J. Sci.,* **37** : 313–319.

14. C. Gabel 1963, Further Human Remains from the Central African Later Stone Age, *S. Afr. archaeol. Bull.,* **18** : 40–48.

15. L. H. Wells 1950, *in* J. D. Clark 1950 : 143–150.

16. F. B. Macrae 1926, The Stone Age in Northern Rhodesia, *Nada,* **4** : 67–68. L. H. Wells 1957, Late Stone Age Human Types in Central Africa, *Proc. Third Panafrican Prehistory Congress,* London : 183–185. J. D. Clark 1942, Further Excavations (1939) at the Mumbwa Caves, Northern Rhodesia, *Trans. roy. Soc. S. Afr.,* **29** : 133–201.

17. University of the Witwatersrand, Johannesburg, Medical School, Hospital Street, Johannesburg, South Africa. Reg. No. 6863.

18. ————————

NACHIKUFU   12° 14′ S, 31° 09′ E.

In the Nachikufu cave and rock shelter, 1 km W of the Great North Road, and 54 km S of Mpika, Northern Province, J. D. Clark and L. Hodges excavated hominid remains from two levels.

Level IIb yielded one upper lt P, and an extant fauna dated 4830 ± 320 BP (Y-319B).

Level III yielded a lower rt M2 (f), upper rt I1 and P3 or P4. The fauna is extant; dates suggest an age not more than 1200 BP

J. D. Clark 1950a, The Newly-discovered Nachikufu Culture of Northern Rhodesia and the Possible Origin of Certain Elements of the South African Smithfield Culture, *S. Afr. archaeol. Bull.,* **5** : 86–98. J. D. Clark 1950, *Stone Age Cultures of Northern Rhodesia,* Cape Town: 137–150. S. F. Miller 1971, The Age of the Nachikufu Industries in Zambia, *S. Afr. archaeol. Bull.,* **26** : 143–146. S. F. Miller 1972, Archaeological Sequence of the Zambian Later Stone Age, *Proc. 6th Pan-Afr. Congr. Prehist.,* Dakar 1967.

---

*R. A. Dart and N. del Grande (1931) record the recovery of fragments of at least 16 individuals but the remainder have not been preserved.   17 teeth were excavated by J. D. Clarke in 1939 from Holocene levels.

'RHODESIAN MAN'

See under **BROKEN HILL**

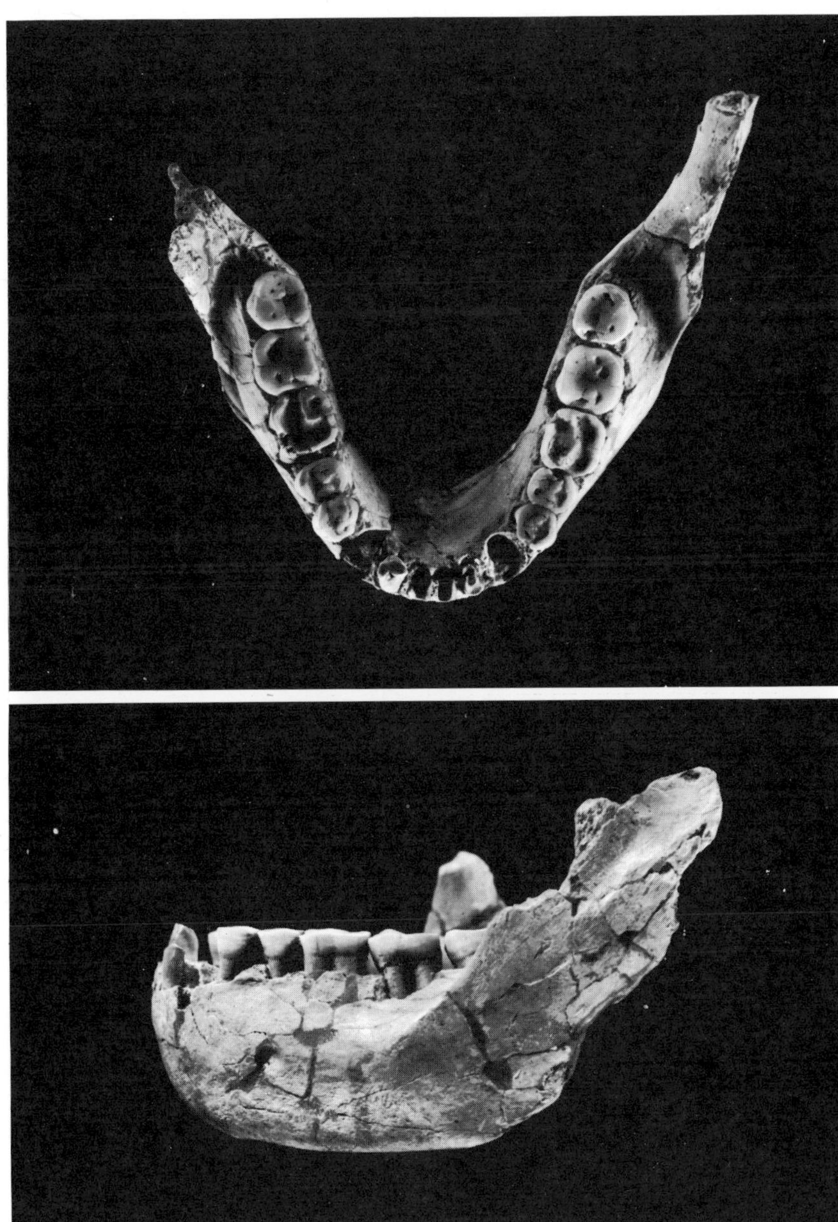

TERNIFINE 1 MANDIBULA

Holotype of *Atlanthropus mauritanicus* Arambourg, 1954. Courtesy of Musée de l'Homme.

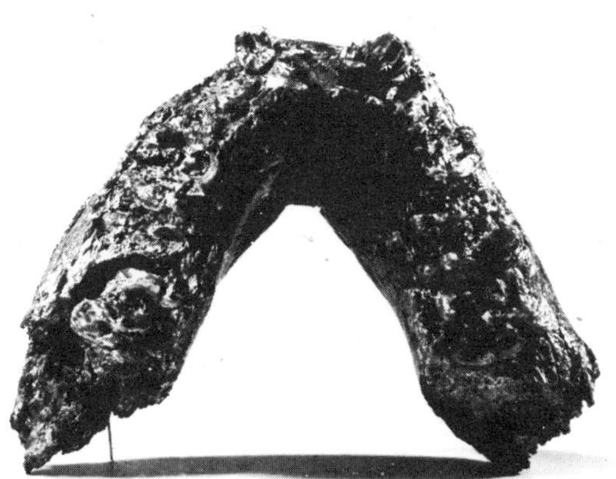

YAYO 1 FRONTO-FACIAL FRAGMENT

Holotype of *Tchadanthropus uxoris* Coppens, 1965. Courtesy of Musée de l'Homme.

OMO 18 CORPUS MANDIBULAE

Holotype of *Paraustralopithecus aethiopicus* Arambourg & Coppens, 1968. Courtesy of Musée de l'Homme.

FORT TERNAN FT 46 AND 47 MAXILLAE FRAGMENTS

Holotype of *Kenyapithecus wickeri* Leakey, 1962. Courtesy of P. Andrews.

KANAM 1 CORPUS MANDIBULAE

Holotype of *Homo kanamensis* Leakey, 1935. BM (NH).

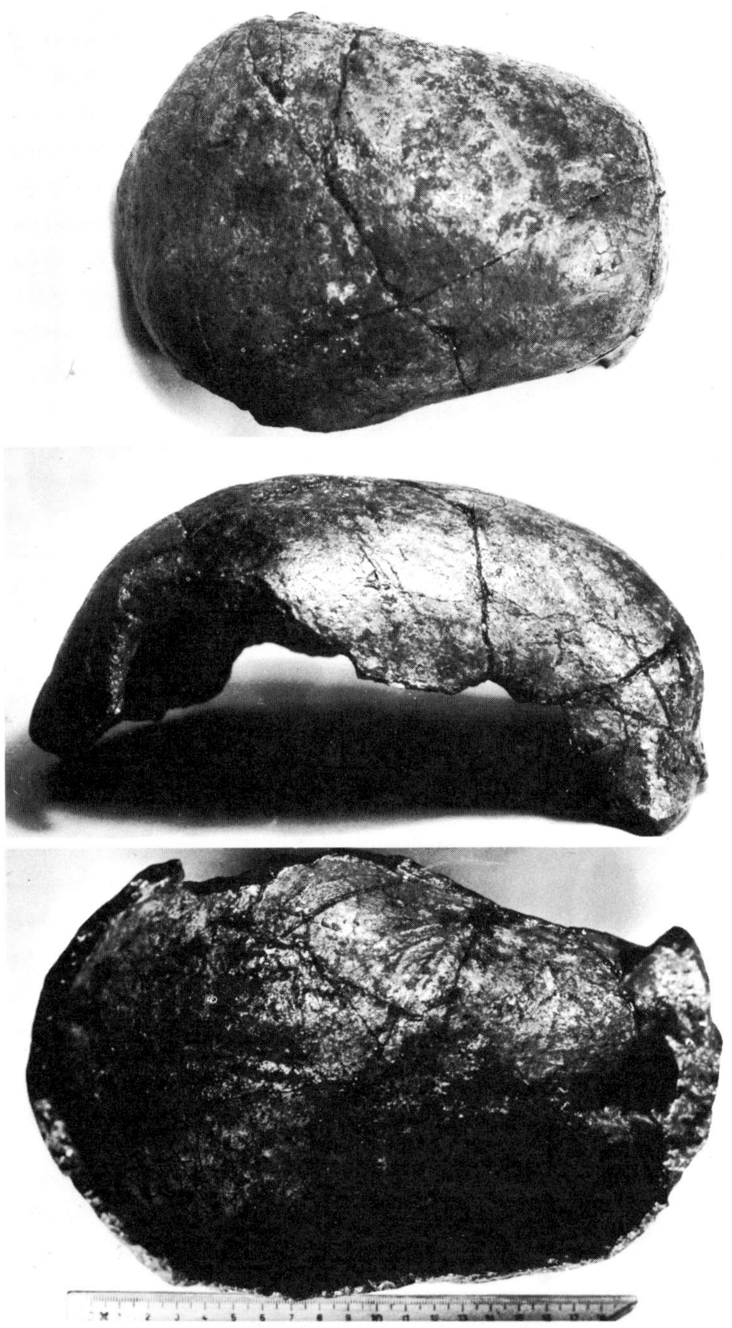

BOSKOP 1 CALOTTE

Holotype of *Homo capensis* Broom, 1917. Courtesy of Port Elizabeth Museum S.A.

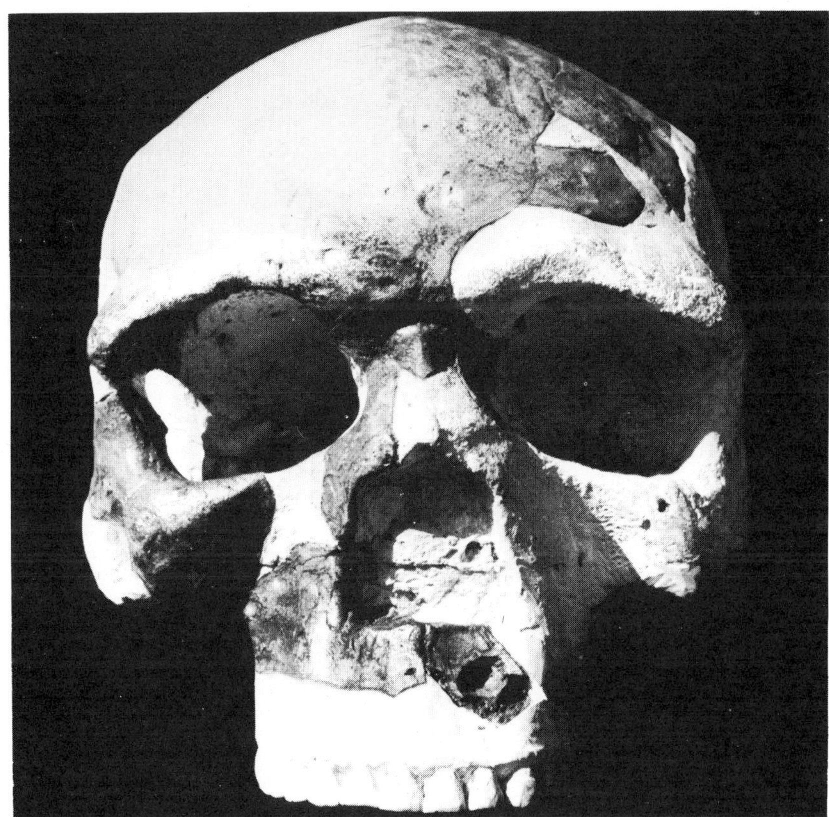

FLORISBAD 1 RECONSTRUCTED CRANIUM

Holotype of *Homo (Africanthropus) helmei* Dreyer, 1935. Courtesy of Nasionale Museum Bloemfontein.

HOPEFIELD 1 CALVARIA

Holotype of *Homo saldanensis* Drennan, 1955. Courtesy of South African Museum.

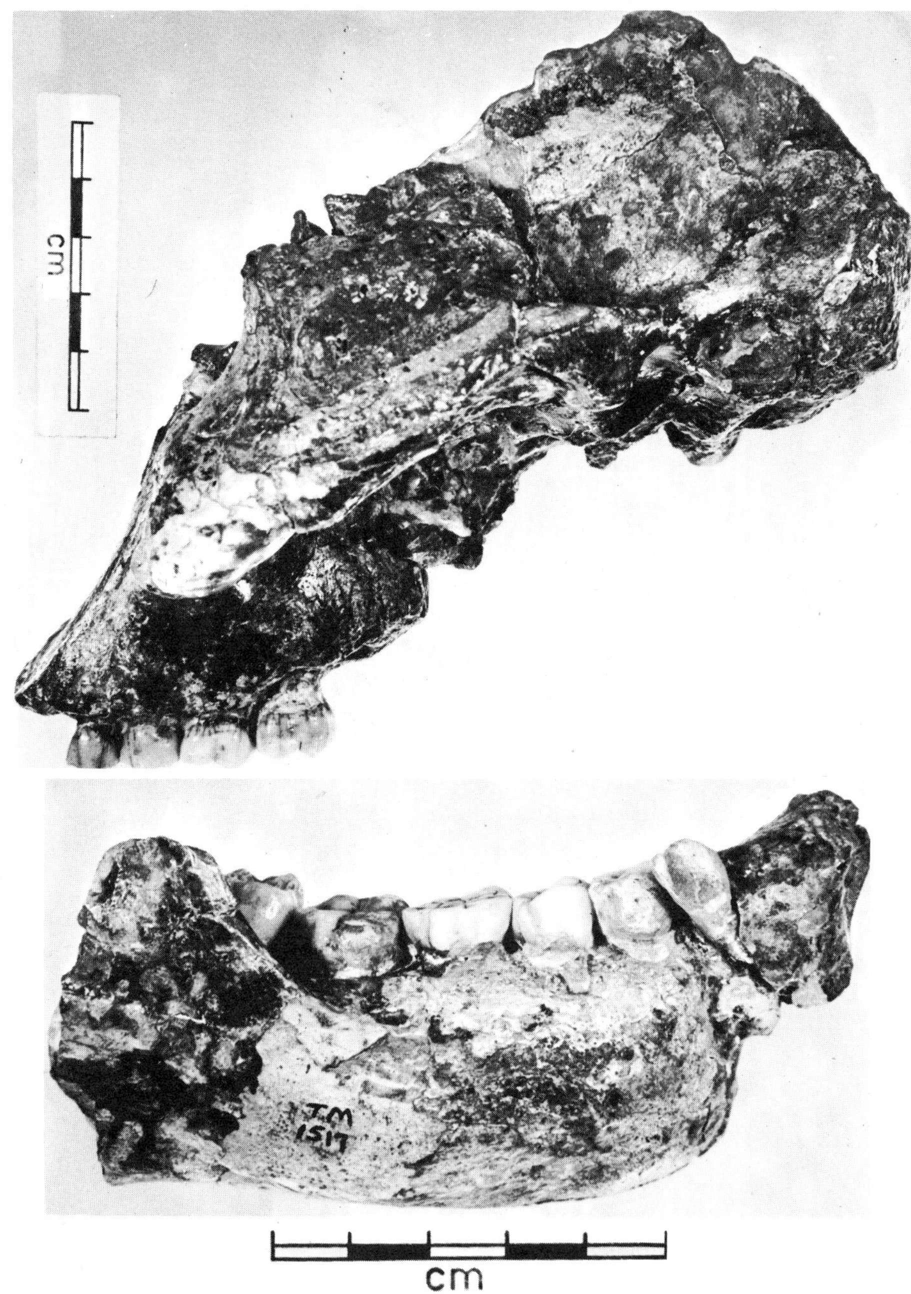

KROMDRAAI TM 1517 CRANIUM AND MANDIBULA

Holotype of *Paranthropus robustus* Broom, 1938. Courtesy of Transvaal Museum, Pretoria.

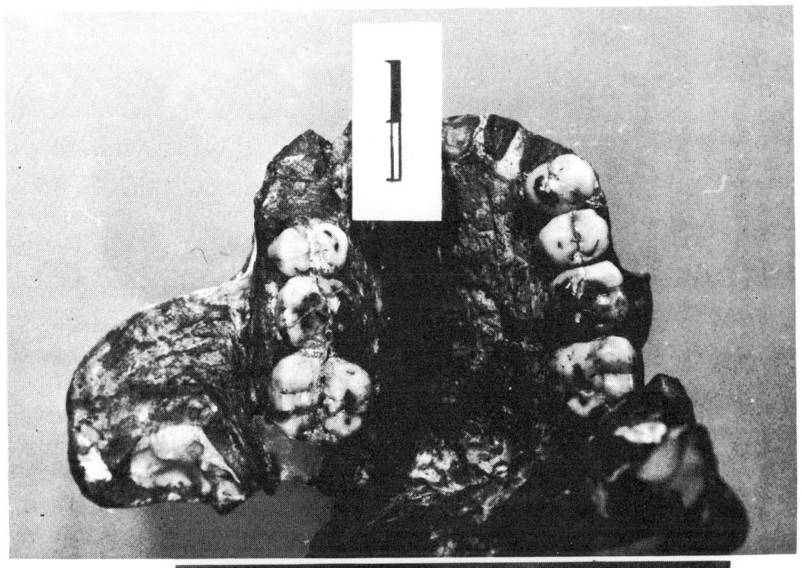

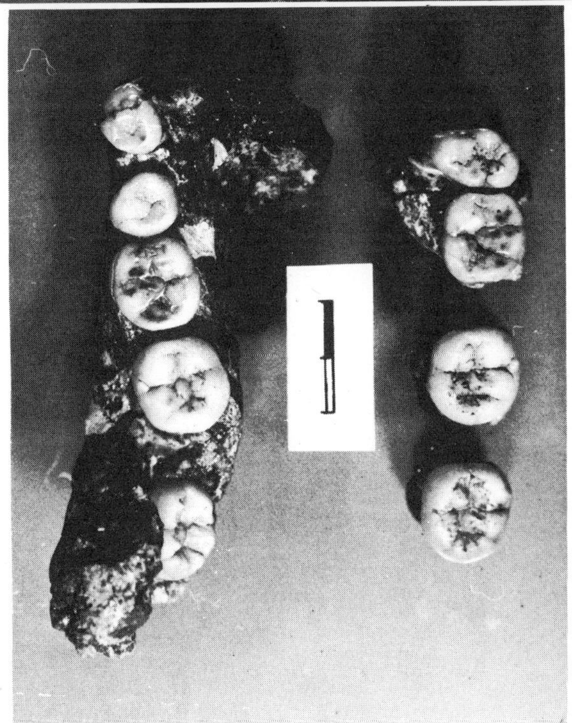

STERKFONTEIN TM 1511 CRANIUM–PALATINUM

Holotype of *Australopithecus transvaalensis* Broom, 1936. *Plesianthropus transvaalensis* Broom, 1938. Courtesy of Transvaal Museum, Pretoria.

SWARTKRANS SK6 MANDIBULA

Holotype of *Paranthropus crassidens* Broom, 1949. Courtesy of Transvaal Museum, Pretoria.

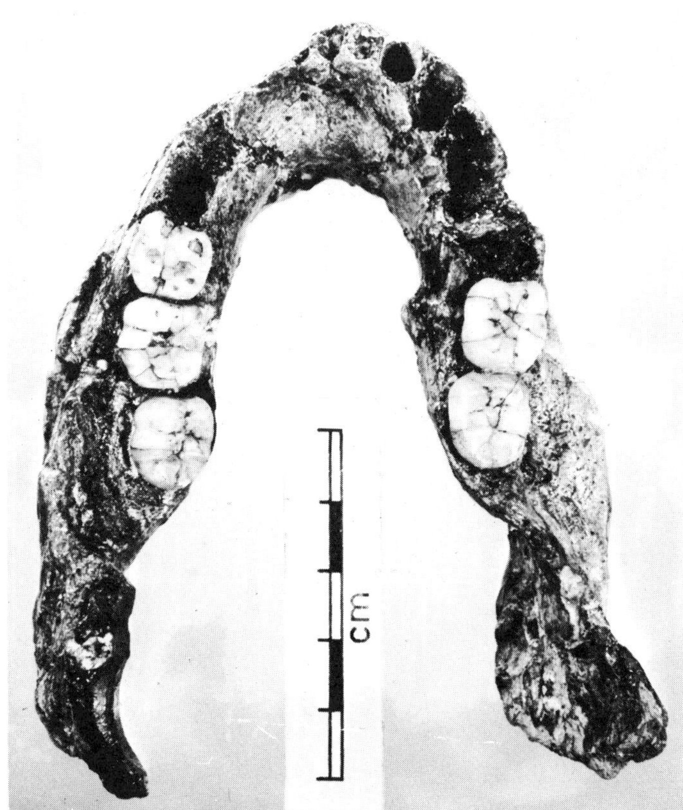

MAKAPANSGAT LIMEWORKS MLD 1 CALVARIA FRAGMENT

Holotype of *Australopithecus prometheus* Dart, 1948. Courtesy of Alun Hughes.

SWARTKRANS SK 15 MANDIBULA

Holotype of *Telanthropus capensis* Broom & Robinson, 1949. Courtesy of Transvaal Museum, Pretoria.

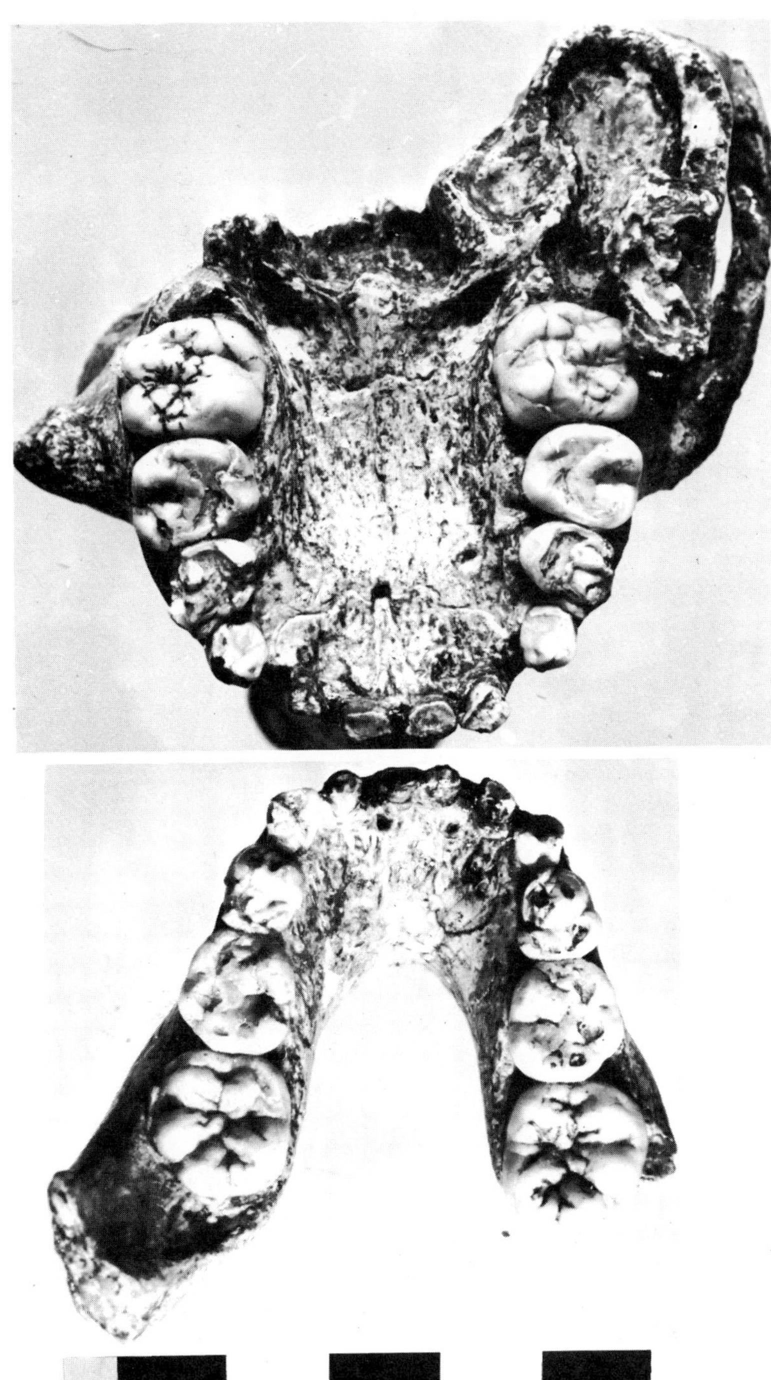

TAUNG 1 SKULL–DENTITION

Holotype of *Australopithecus africanus* Dart, 1925. Courtesy of Alun Hughes.

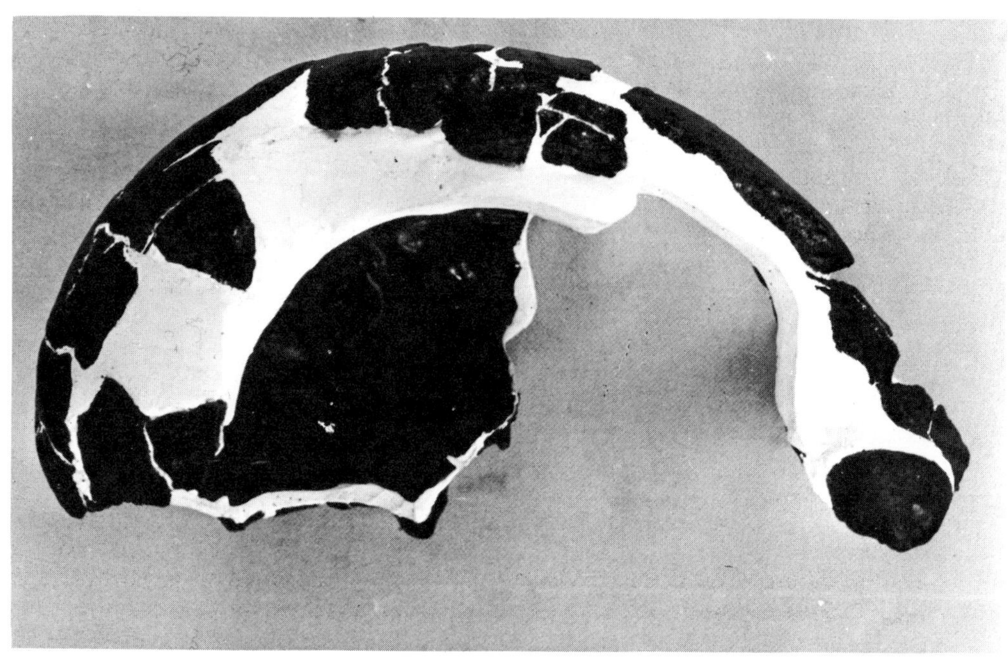

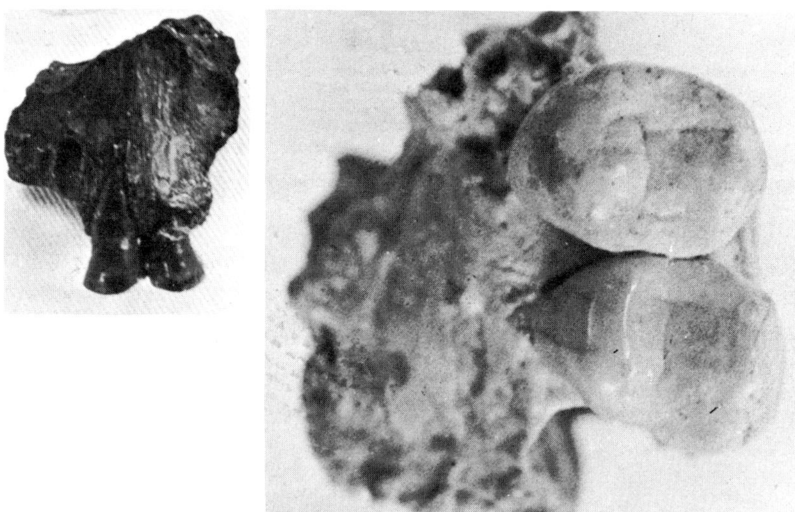

EYASI 1 CALOTTE

Holotype of *Palaeoanthropus njarasensis* Reck & Kohl-Larsen, 1936. Courtesy of Reiner Protsch.

GARUSI 1 RIGHT MAXILLA

Holotype of *Meganthropus africanus* Weinert, 1950. Courtesy of Reiner Protsch.

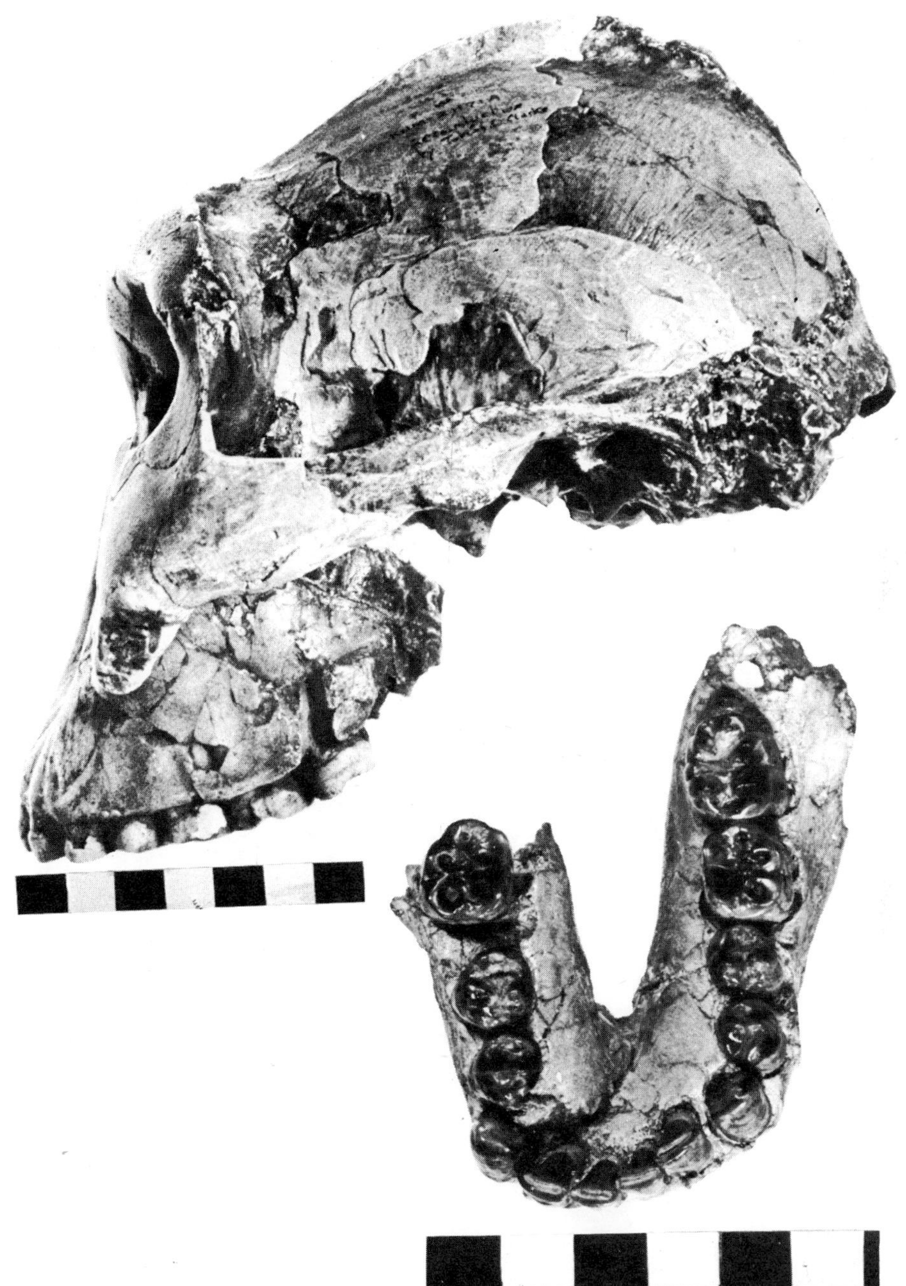

OLDUVAI GORGE OH 5 CRANIUM

Holotype of *Zinjanthropus boisei* Leakey, 1959. Courtesy of National Museum, Nairobi.

OLDUVAI GORGE OH 7 MANDIBULA

Part holotype of *Homo habilis* Leakey, Tobias & Napier, 1964.   Courtesy of National Museum, Nairobi.

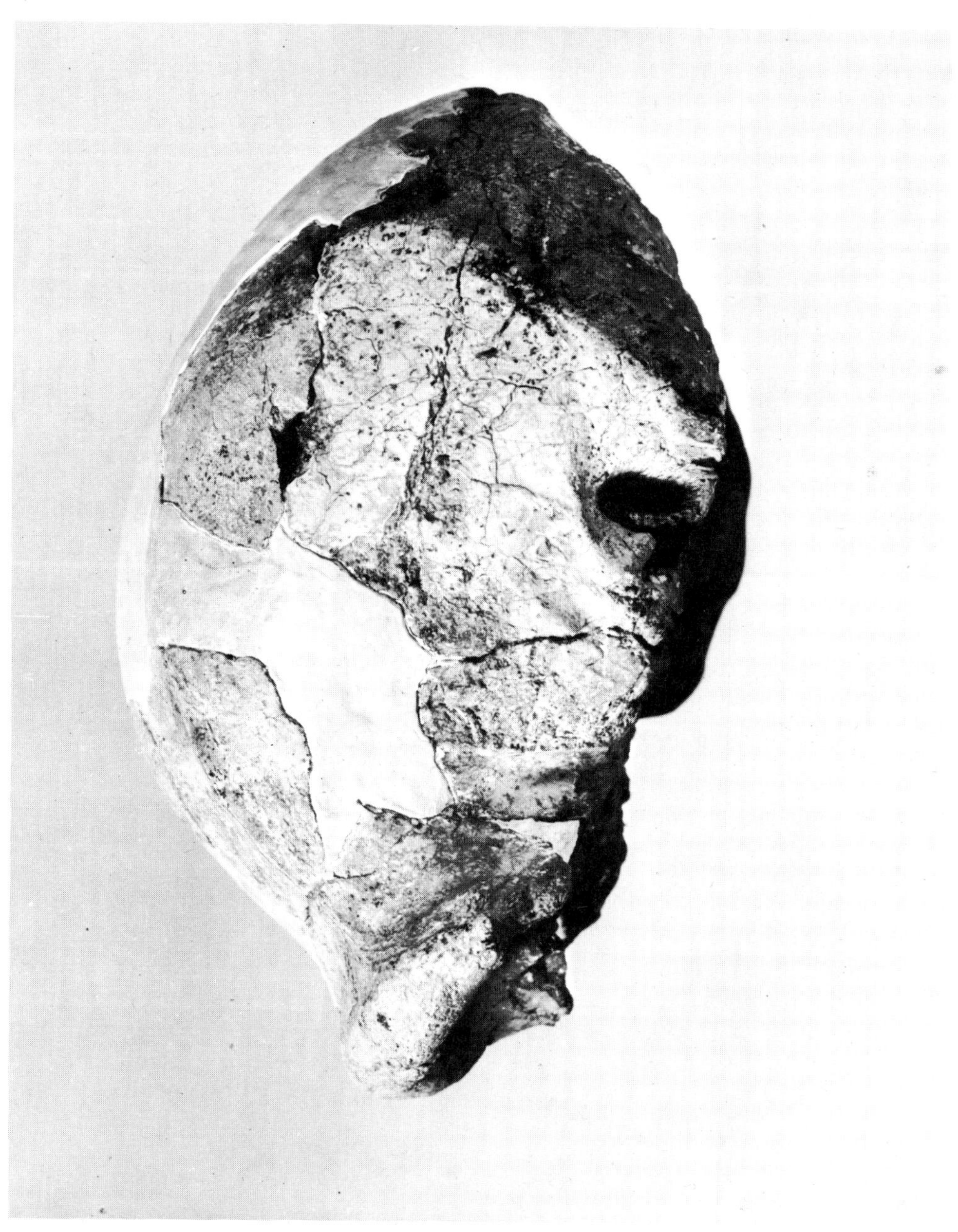

OLDUVAI GORGE OH 9 CALVARIA

Holotype of *Homo leakeyi* Heberer, 1963. Courtesy of Alun Hughes.

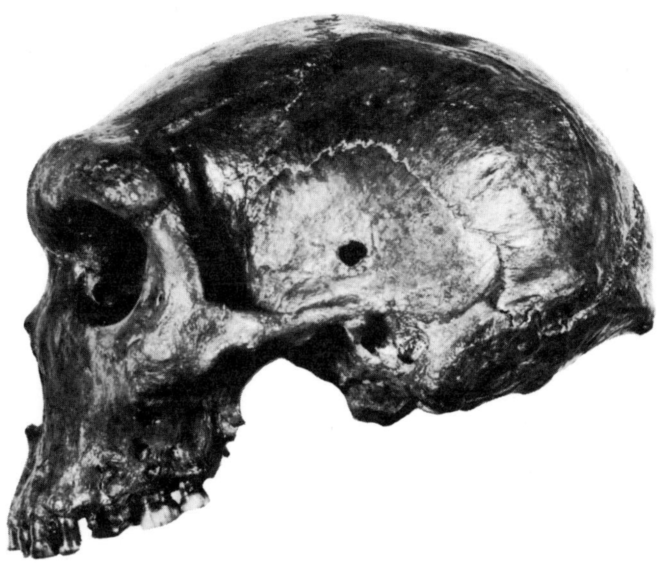

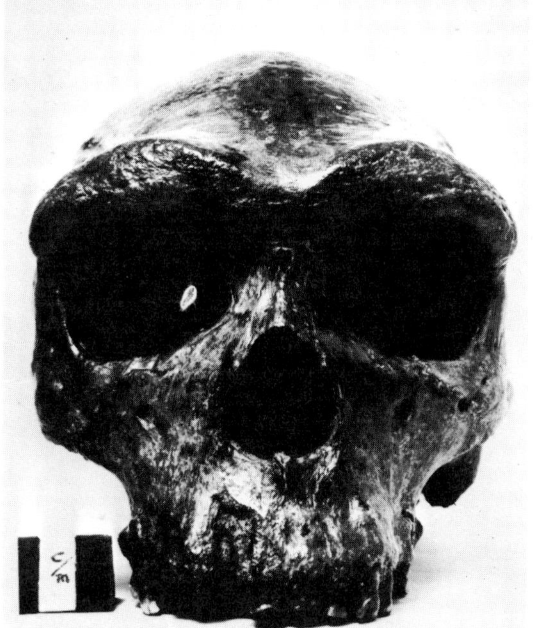

BROKEN HILL 1 CRANIUM

Holotype of *Homo rhodesiensis* Woodward, 1921. BM (NH).

Acetabulum 106, 147
Achatina 74
Achegour 90
Acheulian 21, 32, 83, 103, 166, 181, 182, 183
    A. Final 100
    A. Inferred 181
    A. Lower 186
    A. Moroccan 5
    A. Primitive 5
    A. Upper 87
    A. Upper Middle 84
Acromial extremity 114
Adolescent 3, 61, 83, 86, 100, 112, 113, 114, 124, 137, 138, 140
Adult 3, 4, 6, 8, 10, 11, 13, 15, 18, 20, 22, 23, 32, 34, 37, 46, 50, 58, 61, 64, 70, 71, 74, 78, 80, 81, 82, 84, 85, 86, 88, 92, 96, 97, 99, 101, 102, 103, 104, 105, 109, 110, 111, 112, 113, 117, 122, 123, 124, 125, 126, 127, 129, 131, 133, 138, 139, 141, 144, 147, 148, 150, 154, 156, 158, 165, 178, 180, 184, 186, 188, 192, 194, 195, 196
Aepyceros melampus 63, 98
Afalou-bou-Rhummel 3
Africa, South 95–151
    A. South West 153–154
Africanthropus 163
Afrochoerus 5
    A. nicoli 60, 181
Ain Bahir 2
Ain Beida 2
Aioun Beriche 2
Alain 2
Alcelephas 13, 68, 83
    A. bubalis 84, 88
    A. buselaphus 10, 63
    A. helmei 100, 102
    A. howardi 186
Algeria 1–6
Ali Bacha 2
Aliya *See under* Mugharet el-'Aliya
Alveolar ridge 112
Alveoli 32, 36, 45, 126, 162
Alveolus 139
Amirian 88
    A. Lower 5
Ammotragus 68
    A. lervia 3, 13
Amphibia 166
Ancylotherium 164
Anfabian 84
Antelope 12, 18
Antidorcas 179

A. cf. australis 134
A. bondi 105, 106, 134
A. cf. marsupialis 134
A. cf. recki 105, 106, 134
Archidiskon exoptatus 163
Arrow-heads 10
Arthropoda 63
Artiodactyla 164
Asselar 74
Aterian 82
Aterian, Late 85
Atlanthropus mauritanicus 6, 88
Atlas 42, 46, 94, 188
Auditory meatus 123
Auditory region 138
Australopithecus 33, 43, 44, 46, 47, 48, 49, 50, 51, 52, 53, 54, 56
    A. africanus 119, 149, 164
    A. a. transvaalensis 111, 119
    A. boisei 178
    A. prometheus 110, 111, 112, 113, 114
    A. robustus 149
    A. transvaalensis 125, 126
    A. (Zinjanthropus) 186
    A. (Z.) boisei 172
Aves 12, 63, 88, 156, 164, 166, 188
Axis 14, 94, 144

Ballana 10, 15
Barbus 10
Baringo 32, 33
Base metacarpalis 174
Basis cranii 120, 123, 133, 149
Bekkaria 2
Biparieto-occipitale 109, 110
Bir Oum Ali 2
Bones, cranial 70
    B. post-cranial 34, 70, 115, 144, 147, 150, 154, 192
Border Cave 96
Bos 34, 83
    B. primigenius 3, 5, 10, 12, 13, 81, 84, 88
Boskop 97, 150
Bovidae 24, 158
Brachypotherium 65
Braincase 138
Breccia cast 109, 110, 112, 114
Broken Hill 22, 192
Bromheads Site *See under* Elmenteita
Bubalis cf. nilssoni 162
Burials 3, 4, 5, 10, 11, 15, 18, 34, 61, 70, 71, 86, 92, 96, 99, 101, 104, 115, 117, 150, 156, 184, 194, 195, 198

Bushman Rock Shelter 98
Byneskranskop 99

Calcaneus 37, 61, 96, 116, 189
Calotte 5, 11, 34, 80, 83, 92, 97, 136, 138, 143, 149, 150, 170, 178, 180, 182
Calvaria 5, 10, 12, 22, 23, 41, 48, 71, 81, 84, 103, 109, 110, 120, 121, 123, 133, 157, 170, 179, 199
Camelid 24
Camp site 10
Canidae 63
Canis anthus 81, 88
  C. brevirostris 119
Cape Flats 99
Capitatum 20, 174
Capsian 2, 36, 186
  C. Upper 4, 64
Cap Tenes 2
Caput femoris 61
Caput radii 78
Carnivora 24
Carpi 22
Casablanca Man 85
Cave 2, 3, 18, 68, 80, 81, 84, 85, 86, 87, 96, 99, 100, 101, 103, 107, 117, 118, 134, 145, 192, 193, 197, 198, 199
Cave of Hearths 100
Cemeteries 14, 159
Cephelophus cf. harveyi 32
Ceratotherium 33, 164
  C. simum 58, 84, 180
Cercopithecinae 164
Cercopithecoidea 24, 63, 164
Cercopithecoides williamsi 108, 119, 149
  C. williamsoni 105
Chad 7, 8
Chalicotheridae 24
Champlain 2
Chellean 60
Chemeron *See* Baringo
Chemoigut 33
Chesowanja *See* Baringo
Child 3, 82, 149, 189
Children 20, 70, 86, 117, 156
Chilotheridium pattersoni 65
Chipongwe 193
Chiropterans 24
Clarias 10, 13, 33, 177, 183
Clavicula 22, 70, 71, 96, 109, 110, 114, 171, 174, 199
Colobinae 24, 164
Colobus 22, 32, 188

Columnata 2
Connochaetes 84
  C. cf. africanus 179
  C. taurinus 63, 98, 105, 106, 134
Corpus mandibulae 18, 20, 26, 27, 35, 36, 43, 44, 45, 46, 47, 48, 50, 51, 52, 53, 54, 55, 56, 57, 58, 61, 62, 68, 87, 100, 126, 127, 139, 140, 146, 165, 176, 177, 180, 183
Costa 60, 70, 174
Costae 20, 22, 61, 71, 80, 96, 131
Crania 3, 34, 115, 138, 150, 159
Cranium 3, 4, 11, 33, 37, 41, 42, 43, 48, 51, 53, 54, 56, 57, 64, 70, 71, 80, 81, 86, 94, 96, 99, 101, 102, 105, 110, 111, 116, 120, 121, 122, 123, 135, 136, 137, 138, 150, 158, 162, 166, 170, 171, 172, 192, 193, 194, 195, 196, 199
Crocidura taungsensis 149
Crocodylus 14, 74, 171, 175
  C. niloticus 157
Crocuta 63
  C. crocuta 84
  C. c. angella 134
  C. aff. ultra 172
Crown 20, 26, 27, 44, 49, 52, 54, 56, 63, 105, 106, 116, 125, 127, 128, 129, 130, 133, 141, 143, 144, 146, 172
  C. buccal 148
Crypt 27, 51
Cuboideum 188
Cuneiforme 80
Cuvette 5
Cyphanthropus rhodesiensis 192

Damaliscus 134
  D. antiquus 186
  D. cf. dorcas 134
  D. niro 134, 186
Dar-es-Soltan 80
Deinotherium 164
  D. bozasi 33, 58, 59, 63, 171, 173
  D. hobleyi 65
Denen Dora 19
Dens 52, 53, 65, 137, 143, 188
Dentes 6, 20, 22, 26, 27, 28, 29, 32, 33, 35, 36, 37, 43, 44, 45, 46, 47, 48, 49, 50, 51, 52, 53, 54, 55, 56, 57, 58, 62, 68, 80, 81, 82, 83, 84, 85, 87, 88, 92, 94, 96, 100, 104, 105, 106, 109, 110, 111, 112, 113, 116, 120, 121, 122, 123, 124, 125, 126, 127, 128, 129, 130, 131, 133, 135, 136, 137, 138, 139, 140, 141, 142, 143, 144, 145, 146, 148, 149, 154, 158, 159, 162, 163,

165, 170, 171, 172, 174, 175, 176, 177, 178, 179, 180, 181, 182, 183, 185, 188, 189, 194, 198, 199
   D. deciduous 125, 164
   D. mandibulares 141, 142
   D. maxillares 87, 143, 192
   D. permanent 125, 164
Dentition, complete 126, 172, 180, 186
   D. full 124
Diceros 164
   D. bicornis 102
Dinofelis barlowi 108, 119
Dire-Dawa 18
Dropithecus cf. africanus 35
   D. cf. nyanzae 35

East Rudolf *See* Ileret and Koobi Fora
Ectopotamochoerus dubius 173
Egypt 9–15
Elandsfontein *See* Hopefield
Elephas africanus 10
   E. atlanticus 83
   E. ekorensis 20, 59
   E. iolensis 88
   E. loxodonta 22
   E. recki 20, 22, 28, 33, 60, 171, 173, 175, 176, 180, 181, 186
Elmenteita 34
Endocast 109, 110, 137
Endocranial cast 84, 121, 122, 145
Enhydriodon cf. iluecai 63
Epi-Ouljian 87
Epi-Pleistocene 188
Equidae 24
Equus 33, 68
   E. asinus 12, 13
   E. burchelli 63, 98, 102
   E. b. mauritanicus 82
   E. capensis 101
   E. cf. harrisi 94
   E. helmei 100, 102, 108, 119
   E. kuhni 198
   E. mauritanicus 5, 81, 83, 84, 88
   E. cf. plicatus 103
Escargotière 2
Esna 10
Etheria elliptica 62
Ethiopia 17–29
Eyasi 162

Face 123, 138, 140, 141, 143
Fakhourian 10
Fauresmith 100

Fayum 10, 11, 12
Felidae 63, 164
Female 3, 4, 11, 20, 64, 68, 71, 86, 88, 96, 106, 110, 111, 112, 114, 117, 122, 123, 124, 131, 138, 139, 140, 162, 192, 194, 195, 196
Femora 10, 14, 37, 56, 60, 70, 71, 188, 199
Femur 20, 22, 41, 42, 43, 46, 50, 57, 60, 61, 97, 99, 109, 114, 120, 121, 132, 136, 137, 146, 170
   F. diaphysis 27, 43, 47, 50, 54, 55, 57, 131, 181, 182, 193
   F. d. distal 54
   F. distal 20, 43, 49, 50, 57, 131, 132, 193
   F. neck 48, 132, 178
   F. proximal 20, 48, 49, 52, 54, 57, 116, 132, 144, 178, 193
Fibula 3, 22, 47, 61, 70, 71, 80, 97, 170, 172, 199
   F. distal 20, 56, 57
Fibulae 47, 199
Fingira 70
Fish Hoek 101
Florisbad 102
Foot bones 170
Footprints 76, 115
Foramen magnum 123
Fort Gouraud 76
Fort Ternan 35
Fragments, cranial 20, 41, 42, 44, 46, 47, 49, 51, 52, 53, 56, 57, 61, 125, 138, 170, 178, 188, 194
   F. cranio-facial 110
   F. fronto-facial 8, 23
   F. ischial 109, 113
   F. mandibular 20, 49, 61
   F. maxillary 20, 145
   F. post-cranial 47, 61, 196
   F. shaft 61
Frontale 13, 14, 27, 60, 64, 88, 96, 109, 110, 123, 162, 170, 180, 188

Galagines 24
Galago senegalensis 172
Gambetta 2
Gamble's Cave 36
Gamblian 162
   G. Upper 63
Garusi 163
Gazella 68, 74, 83, 84, 105, 106
   G. atlantica 81, 88
   G. bondi 102
   G. dorcas 3, 13, 82

G. isabella 14
G. rufifrons 10
G. vanhoepeni 134
G. wellsi 172, 173
Gigantohyrax 108
Giraffa camelopardalis 58, 82
  G. cf. jumae 63, 164
Giraffidae 24
Glenoid fossa 138
Gombore 21
Gomphotherium 35
Gorgon olduvaiensis 173
Grave 196
  G. shafts 14
Guomde Formation 37, 45
Gwisho 193, 194

Hadar 19
Hammerstones 165
Hand bones 170, 171
Haua Fteah 68
Helix 3
Hexaprotodont 20
Hipparion 20, 164
  H. cf. libycum 108
  H. turkanese 63
Hippopotamidae 24
Hippopotamus 8, 12, 63
  H. amphibius 8, 13, 32, 33, 82, 88, 102, 103,
    188
  H. gorgops 176, 177, 183
  H. imaguncula 58
Hippotragus cf. niger 134
  H. cf. Kobus ellipsiprymnus 134
Holocene 2, 3, 4, 5, 12, 34, 36, 61, 64, 70, 71,
    74, 90, 99, 115, 117, 150, 154, 156, 194,
    195, 196, 197, 198
Homo 50, 51, 52, 54, 55, 56, 58, 138, 171
  H. africanus 111
  H. australoideus africanus 99
  H. capensis 97
  H. drennani 99
  H. erectus 21, 32, 52, 147, 166, 182
  H. habilis 172, 173, 174, 176, 177
  H. kanamensis 58
  H. leakeyi 179
  H. neanderthalensis 81, 82
  H. rhodesiensis 192, 193
  H. saldanensis 103
  H. sapiens 3, 22, 23, 34, 37, 58, 64, 71, 97,
    102, 158
  H. s. afer. 64
Homoioceras 162

H. baini 100, 102, 103, 150, 192
  H. singae 157
Hopefield 103
Hora 71
Horse 18
Humeri 14, 20, 22, 70, 71, 199
Humerus 41, 42, 43, 49, 50, 61, 71, 109, 110,
    121, 136, 188, 199
  H. diaphysis 26, 44, 114
  H. distal 43, 46, 53, 54, 59, 105, 114, 132,
    144, 193
  H. proximal 57, 126
Hyaena brunnea 102
  H. crocuta spelaea 88
  H. spelaea 82
Hyaenidae 164
Hyracoidea 24
Hystrix 164
  H. astasobae 157
  H. cristata 3, 88

Ibero Maurusian 2, 3, 5, 80, 86
Ileret and Koobi Fora 37–57
Ilia 14, 96
Ilium 61, 106, 109, 110, 113, 114, 132, 137,
    147, 192
Infant 3, 15, 70, 81, 96, 98, 104, 110, 114, 117,
    146, 195, 196
Interpluvial, First 119
Inyanga 94
Ischium 61, 96, 193
Ishango 188
Iwo Eleru 92

Jebel Irhoud 80
Jebel Silsila 12, 13
Jebel Taya 2
Juvenile 3, 10, 20, 26, 34, 43, 46, 47, 49, 50,
    51, 52, 55, 99, 106, 112, 124, 125, 126,
    133, 138, 139, 140, 141, 144, 165, 173,
    180, 188, 195, 196

Kabua 58
Kabwe Man *See under* Broken Hill
Kada Hadar 19
Kakontwe 189
Kalemba 196
Kanam 58
Kanapoi 59
Kanjera 60
Kapthurin *See under* Baringo
Kassimatis 78
Kem Oum Touiza 2

Kenya 31–65
Kenyapithecus wickeri 36
Khanguet el Mouhaad 2
Khartoum 156
Kibish 22
Klasies River Mouth 103, 104
Klipfonteinrand 104
Kobus sigmoidalis 172
   K. ellipsiprymnus 134
   K. venterae 102
Kom Ombo 12, 13
Koobi Fora *See under* Ileret and Koobi Fora
Koro Toro 8
Kromdraai 105, 106

Laetolil 164
Lagomorph 24
Lake Besaka 18
Lambda 110
La Mouillah 3, 4
La Tranchee 2
Lanistes carinatus 62
Lates niloticus 13, 74
Le Chacal 2
Leakeymys ternani 35
Leopard's Hill 197
Leporid 158
Lepus capensis 10
   L. kabylicus 88
Levalloisian 157, 158, 162
Levalloiso-Mousterian 65, 81
Libya 67, 68
Libytherium 108, 186
   L. olduvaiensis 58, 176, 177, 179, 180
Limeworks 110, 111, 112, 113, 114, 148
Limnopithecus legetet 35
Litocranius cf. walleri 192
Loboi 61
Lochinvar *See under* Gwisho
Long bone 41, 42, 92, 96
Lothagam 62
Loxodonta 164
   L. adaurora 20, 59
   L. africana 39, 43, 44, 45, 46, 47, 49
   L. atlantica 6, 8, 103
Lukeino 62
Lukenya Hill 63
Lutra 175
   L. maculicollis 188
Lycyaena 119
   L. silberbergi 134

Machairodont 164
Machairodus 6
Madoqua 164
Magosian 99
Maiko Gully 185
Makalian 188
Makapania 134
   M. broomi 108, 119
Makapansgat *See under* Cave of Hearths
   M. limeworks 107–114
   M. skull 111
Makwe 196, 197
Malar 36, 82
Malawi 69–71
Male 3, 4, 6, 13, 37, 50, 61, 64, 70, 71, 74, 80, 81, 83, 85, 86, 92, 96, 105, 112, 113, 117, 122, 124, 126, 138, 139, 144, 154, 162, 172, 180, 184, 192, 194, 195
Mali 73
Mammalia 24, 157, 166
Mandible 41, 42, 43, 170
Mandibula 6, 10, 11, 22, 26, 32, 37, 56, 57, 64, 70, 71, 81, 83, 85, 92, 94, 96, 97, 98, 99, 101, 104, 105, 106, 109, 110, 112, 113, 116, 120, 121, 124, 126, 135, 136, 139, 140, 141, 142, 145, 146, 148, 149, 150, 154, 158, 173, 186, 188, 194, 195, 196, 199
Mandibulae 34, 115, 139, 140
Manus 3, 71
Maramba 198
Marnia 4
Masek Beds 165
Mastoid 144
Mastoideus 137
Matjes River 114
Mauritania 75
Maxilla 6, 20, 36, 37, 41, 42, 47, 51, 55, 82, 83, 84, 88, 105, 106, 109, 110, 112, 116, 120, 121, 122, 123, 124, 125, 129, 133, 135, 136, 137, 138, 140, 146, 162, 163, 192, 195
Maxillae 22, 111, 122, 124, 138, 139, 140, 199
Meatus acusticus externus 138
Mechta Chateaudun 4
Mechta-el-Arbi 2, 4, 5
Megalotragus 134
Meganthropus africanus 163
Melka Kontoure 21
Mesloug 2
Mesochoerus lagetani 103
   M. limnetes 39, 49, 56, 57
   M. olduvaiensis 176, 180

Mesolithic 156
  M. Early 188
Metacarpalia 20, 22, 80, 188, 189
Metacarpale 105, 121, 122, 132, 133, 136, 137,
  147, 174
Metatarsale 32, 41, 42, 46, 47, 51, 56, 170
Metatarsalia 22, 37, 80, 174, 188
Metridiochoerus andrewsi 39, 43, 45, 48, 49, 50,
  51, 52, 186
Microlithic 12
Miocene, Middle 35
  M. Late 63
Moçambique 77, 78
Mollusca 10, 63, 84, 156, 166, 188
Morocco 79–88
Moruarot Hill 64
Mugharet-el-'Aliya 81
Mumbwa 198
Mytilus 3

Nachikufan 196, 197
Nachikufu 199
Nahoon Point 115
Naisiusiu Beds 170, 184
Naivasha 64
Nasalia 122
Natron *See under* Peninj
Naviculare 61
Ndutu 165
Neandertaloid 18
Negroid 74
Nelson Bay 116
Neumours 4
Neolithic 10, 11, 74
Nesokia indicia 10
Ngorora 65
Niger 89, 90
Nigeria 91, 92
Notochoerus 108, 164
  N. capensis 28
  N. euilus 20
Nyanzachoerus 28
  N. pattersoni 20, 59
  N. plicatus 59
  N. tulotos 63

Oakhurst 116
Occipitale 11, 13, 26, 37, 42, 49, 60, 62, 96,
  110, 122, 123, 138, 162, 170, 174, 197
Occipitalia 14
Occlusal caries 140
Oiceros tanyceras 35
Okapia stillei 163

Old Kingdom 11
Oldowan 24, 119, 134, 166, 172, 173, 178,
  180
  O. Developed 176, 177, 179
  O. Inferred 171, 175
Olduvai Gorge 166–185
Omo 22–29
Open site 10, 11, 13, 15, 21, 22, 37, 97, 102,
  103, 154, 156, 157, 164, 165
Oranian 2, 3, 5, 86
Orbita 88, 124
Oreotragus cf. major 134
  O. cf. oreotragus 134
Orycteropus 63, 164
Os capitatum 120, 131
Os coxae 20, 43, 50, 120, 121, 131, 132, 136,
  144, 170, 182
Os cuboideum 61
Os metacarpalia 96
Ossa cuneiforme 61
Ossa faciei 105, 120, 121, 123, 124, 133, 136,
  138, 149
Ossa longa 64, 195
Ossa mani 3
Ossa metatarsalia 61, 146
Ossa tarsi 174
Ossuary 3, 4, 86
Ostrea margaritacea 78
Otjiseva 154
Otocyon recki 173
Oued Medfoun 2
Ouljian 82
Ourebia cf. ourebi 134

Palaeoanthropus njavasensis 162
Palaeolithic, Late 10, 15
  P. Middle 18
Palaeotragus 65
  P. primaevus 35
Palate 41, 123
Palatinum 52, 105, 106, 109, 112, 120, 121,
  123, 125, 133, 137, 138, 139, 140, 145,
  146, 170, 180, 182, 183
Panthera pardus 188
Papio angusticeps 105, 149
  P. anubis neumanni 34
  P. izodi 149
  P. robinsoni 134
  P. wellsi 149
Papionines 24
Paradiceros mukirii 35
Paranthropus crassidens 137, 138, 145
  P. robustus 105

P. r. crassidens 140, 141, 142, 143, 144
Parapapio antiquus 149
   P. broomi 108, 119
   P. jonesi 59, 108, 119, 149
   P. whitei 108, 119, 149
Paraustralopithecus aethiopicus 26, 27
Parietale 6, 12, 13, 21, 41, 42, 45, 50, 51, 54,
   56, 57, 60, 61, 62, 64, 85, 105, 109, 110,
   111, 123, 138, 162, 170, 172, 174, 180,
   181, 185, 188, 192, 195, 197
Parietalia 11, 14, 26, 96, 110, 123, 173
Parmularius 134
   P. altidens 171, 172, 173
   P. angusticornis 179
Pars mastoidea 110
Pars petrosa 20
Patella 14, 37, 116, 188
Pedetes 164
Peers' Cave *See under* Fish Hoek
Pelea cf. capreolus 134
Pelorovis 104
   P. oldowayensis 60, 176, 180
Pelusios cf. sinuatus 65
Pelvis 10, 70, 71, 137, 147
Peninj 186
Perissodactyla 164
Pes 22, 70, 71, 174
Phacochoerus 32, 74
   P. aethiopicus 63, 98, 102
   P. a. africanus 34
   P. a. mauritanicus 82
   P. africanus 88
   P. antiquus 134
   P. dreyeri 115
Phalanges 20, 22, 41, 47, 80
   P. manus 20, 22, 51, 188, 189
   P. media 173
   P. pedis 188
   P. proximalis 173, 174
Phalanx 32, 60, 116
Phalanx, distalis 173
   P. manus 26, 121, 177
   P. m. proximal 105, 132
   P. pedis 189
   P. p. distal 105, 173
   P. proximal 105
Phenacotragus cf. recki 186
Philippi *See under* Cape Flats
Phoca 85
Pig 18
Pisces 10, 13, 33, 62, 63, 156, 157, 166, 188
Plateles alba 188
Pleistocene 98, 101

P. Late 5, 15, 61, 74, 78, 92, 97, 99, 115,
   150, 157, 158, 159, 184, 196, 197, 198
P. Lower 8, 24, 32, 33, 40, 149, 166, 171,
   172, 173, 175, 176, 177, 178, 179, 180,
   181, 182, 183, 186
P. Middle 5, 6, 8, 21, 22, 32, 60, 105, 106,
   134, 149, 165, 182, 183
P. Upper 10, 12, 13, 14, 15, 18, 22, 96, 100,
   102, 103, 104, 157, 162, 166, 189, 192
Plesianthropus 111
   P. transvaalensis 122, 124, 127, 132
Pliocene 59, 62
   P. Lower 65
   P. Middle 28
   P. Upper 20, 23, 24, 40, 164
Potamochoeroides 108
Potamochoerus 32, 164
   P. intermedius 173
   P. majus 60
   P. porcus 188
Pre-Grimaldian 83
Pre-neandertalian 87
Primelephas gomphotheroides 62, 63
Proboscidea 24, 164
Procavia 34
   P. antiqua 108, 119
   P. transvaalensis 108, 119
Processus coronoideus 61
Processi coracoidei 22
Promesochoerus mukiri 172
Pronotochoerus jacksoni 33
Protragocerus 65
   P. labidotus 35
Proto-Stillbay 157
Pesudocivettictis ingens 173
Pseudotragus 65
Pubis 61, 132, 144

Qadan 15, 156
Qau 14
Quarry 78, 82, 88, 150, 189

Rabat 82
Rabat Man 83
Radii 3, 20, 22, 70, 71, 100, 188
Radius 3, 20, 43, 54, 56, 61, 97, 109, 121, 135,
   171, 188, 198
   R. diaphysis 47, 114, 174
   R. distal 132
   R. proximal 27, 53, 114, 132, 148
Ramapithecus wickeri 36
Rami mandibulae 14, 188, 189

Ramus mandibulae 37, 68, 88, 97, 103, 120, 121, 126, 127, 133, 140, 188, 189
Raphicerus cf. campestris 134
Redunca aff. ancystrocera 62
  R. arundinum 98, 134
  R. darti 108, 119
Reptilia 63, 65, 88, 156, 157, 164, 166
Rhinoceros mercki 81
  R. simus 83, 84
Rhinocerotidae 20, 24
Rhodesia 93
Rhodesian Man *See under* Broken Hill
Ribs 171, 196
Rock shelter 2, 3, 4, 36, 63, 64, 70, 71, 92, 94, 98, 100, 101, 104, 114, 196, 197, 199
Rodentia 24, 164
Rodents 18

Saccostomus 164
Sacrum 20, 192
Sahaba 156
Saldanha *See under* Hopefield
Sale 84
Sandpit 5, 99
Scaphoideum 174
Scapula 20, 61, 70, 71, 96, 120, 126
Sebilian, Lower 12
  S. Middle 12, 13
Setif 4
Shell-midden 4, 116
Shungura 24
  S. Formation 20, 24
Sidi Abderrahman 84
Sidi Hakoma 19
Simopithecus danieli 134
  S. darti 108
  S. oswaldi 60, 173
Singa 156
Sivatherium 28, 164
  S. oldowayensis 60
Skeleton 3, 37, 42, 61, 86, 90, 99, 104, 116, 150, 156, 170, 184, 194, 195, 198
  S. complete 3
  S. post-cranial 3, 10, 74, 101, 194
Skhul 23
Skildergat *See* Fish Hoek
Skull 4, 42, 50, 74, 145, 195
Soleb 157
Soltanian 81, 82
  S. Final 80, 86
  S. Late 85
South Africa 95–151

South West Africa 153–154
Spheroids 165
Springbok Flats *See under* Tuinplaas
Stegotetrabelodon orbus 62, 63
Sterkfontein 118–133
Stillbay 101, 189
Stone Age, Late 64, 70, 71, 92, 94, 98, 99, 104, 116, 117, 193, 194, 195, 196, 197, 198
  S. A., Middle 18, 78, 97, 102, 104, 150, 198
Strandlooper 150
Struthio 12, 88, 177
  S. camelus massoicus 188
Stylohipparion 59, 176, 177
  S. albertensis 58
Sudan 155–159
Suidae 24
Sus 14
Swanscombe 23
Swartkrans 134–148
Symphysial region 50, 51, 58, 112, 126
Symphysis 6, 112
  S. mandibulae 158
Syncerus 134
  S. aff. caffer 22
Synodontis schall 156

Taforalt 85, 86
Talus 20, 41, 42, 43, 47, 51, 61, 96, 105, 188, 189
Tangier Man 82
Tanzania 161–186
Tapinochoerus 119
  T. meadowsi 134
Tarsi 22
Taung 148
Taurotragus oryx 63, 134
Tchadanthropus uxoris 8
Tebessa 2
Teeth 41, 42, 43, 106, 170
Telanthropus capensis 148
Temara 87
Temporale 14, 20, 33, 61, 62, 88, 97, 105, 109, 110, 123, 138, 162
Temporalia 96, 123
Tensiftian, Early 84
Ternifine 5, 6
Testudo 88
Theropithecines 24
Thomas Quarries 88
Thryonomys 32
  T. arkelli 156
  T. swinderianus 188

Tibia 3, 20, 37, 41, 42, 51, 56, 57, 97, 131,
    170, 172, 188, 193
  T. diaphysis 46, 47, 52, 181
  T. distal 49, 56, 116
  T. proximal 20, 43, 46, 54, 57
Tibiae 14, 22, 42, 70, 71, 188, 189
Tilapia 33
Tragelaphus sp. aff. angasi 134
  T. cf. Makapania 134
  T. cf. nakuae 20, 28, 62
  T. cf. scriptus 134
  T. spekei 32
  T. cf. strepsiceros 134
Trapezium 174
Trochus 3
Tsitsikama 150
Tuinplaas 150
Tunisia 186
Tuskha 15
Tympanic plate 110
Tyrrhenian, Post 83

Ulna 3, 20, 22, 32, 47, 56, 61, 71, 97, 170, 180,
    188, 189, 199
  U. proximal 56, 105, 116
Ulnae 3, 20, 70, 188
Ungulata 21
Unio 10
Ursus arctos bibersoni 88
Usno 20, 28

Vertebra lumbalis 27, 144, 147
  V. thoracicalis 116, 147
Vertebrae 41, 61, 70, 71, 96, 120, 121, 136,
    137, 196
  V. centra 131, 132
  V. cervicalis 51, 188
  V. lumbales 20, 22, 131
  V. sacral 131
  V. thoracicae 22
  V. thoracicales 131
Villafranchian, Late 108, 119, 149
Viverridae 164

Wadi Halfa 158
Wilton 94, 117, 193, 194, 195
Würm, Early 65

Xenohystrix sp. 108

Yayo 8

Zaire 187–189
Zambia 191–200
Zinjanthropus boisei 172
Zygoma 22
Zygomaticum 96

35⁰⁰
ф
VG/-

JC

Archaeology